Science 2
for Little Folks

Nancy Nicholson

With utmost gratitude to my patient husband, Marvin, to the entire Johnson family, Mr. John Dunlap, and Brother Scott Nicholson for their encouragement and valuable suggestions, and especially to Sheri Maguire, B.S., M.S., who graciously found time apart from her Biology students and numerous commitments to review these stories.

"You are—and You are the God and Lord of all that You have created: and before Your face stand the causes of all things transient and the changeless principles of all things..."

—St. Augustine: Confessions, I, 6

Copyright © 2003, 2007 Nancy Nicholson

Science 2 *for Little Folks* is under copyright. All rights reserved. No part of this book may be reproduced in any form by any means—electronic, mechanical, or graphic—without prior written permission. Thank you for honoring copyright law.

For Little Folks
P.O. Box 571
Dresden, OH 43821

ISBN: 978-0-9771236-2-6

Cover and interior design by Osprey Design, www.ospreydesign..com

Dear RoseMary, Annie, and Maria;

Who would have guessed, when some of these stories were first written for you, that other children might also be interested in reading them? But then, Our Lord and Lady have so many surprises for us that we should not be surprised!

God shows Himself to us through His Creation; science is just a way of studying and discovering what God already knows, because He is the One Who started it all. When we read stories from science, we learn more about our wonderful God and Father whose might and glory are reflected in all that He created.

Would you like to learn more about what God has made? These stories are for RoseMary, Annie, Maria,

...and YOU!

Contents

1. **Invisible Marks**   1
   Biology: Exocrine glands: Pheromones

2. **How to Dress a Duck**   4
   Zoology

3. **Precious Blood**   7
   Biology: Circulatory system: Blood

4. **Coloring Adam and Eve**   10
   Biology: Skin: Pigmentation

5. **He Will Dry Every Tear**   13
   Biology: Exocrine glands: Lacrimal glands

6. **Our Hearts and Theirs**   16
   Biology: Circulatory system: Heart

7. **Run for Your Life!**   19
   Biology: Exocrine glands: Hormones

Eyes to See   22
Biology: Eyes

Nerves and Your Sense of Touch   25
Biology: Nervous system: Nerves

Ears to Hear   28
Biology: Hearing

Spit!   31
Biology: Digestive system: Saliva

Soul Food   34
Health: Nutrition

Germs Make Me Sick!   37
Biology: Viruses and bacteria

Salt of the Earth   40
Geology: Minerals and solubility

God's Building Blocks   43
Chemistry: Elements, compounds; molecules, atoms

Freezing and Solids   46
Physical science: Freezing and solids

Light from Light   48
Astronomy: Solar system: Sun and moon

Falling Stars   51
Astronomy: Solar system: Meteors

Searching the Heavens   53
Astronomy

From the Rising of the Sun   56
Astronomy: Solar system; Earth's rotation

Merciful Rain   59
Earth Science: Weather: Precipitation

Lightning!   62
Earth Science: Weather: Lightning

Weathering the Storm   64
Earth Science: Geology: Weathering

Rock My Soul   68
Earth Science: Geology: Volcanic, metamorphic rock

Layers of Rock, Layers of Faith   71
Earth Science: Geology: Sedimentary rock, fossils

Feeding Baby Plants   74
Botany: Seeds: Structure and germination

Traveling Seeds   77
Botany: Seeds

So, What's the Difference?   80
Biology: Entomology: Butterflies and moths

Tasty Moth or Scary Owl   83
Biology: Entomology: Moths

It's a Dirty Job   85
Biology: Annelids

Life-giving Water   87
Biology: Habitats and ecosystems

In Wisdom You Have Made Them All   90
Biology: Migration, hibernation, and estivation

Dominion Over the Earth   95
Environmental stewardship

Before You Were Born   98
Biology: Fetology

Baby Food   101
Biology: Infant nutrition

Power in Weakness   104
Biology: God's precious children

Answer Key   108

Key to Pronunciation   111

Sources   113

Biology

Exocrine glands: Pheromones

Invisible Marks

Once we had an old grandma cat named Nip. She showed her affection for us by purring and rubbing her head or arched back against our legs. Sometimes she would rub against doorways and even the chairs where we sat. Have you ever seen a cat do this?

Cats have 'scent glands' on their foreheads, around their mouths, and the lower part of their backs. These contain **pheromones**, which leave a scent that humans can't smell, but other cats can. Have you ever wanted to make sure that your little brother or sister didn't accidentally take something of yours, so you wrote your name on it? You marked it to show your ownership. Since cats can't write, they instead leave an invisible mark on what they think belongs to them, using pheromones.

Sometimes when we visited a friend's house, I liked to pet their friendly cat, which would happily rub against my ankles. When we came home, Nip was always glad to see me, until she smelled my ankles. Then she would hiss and run off to pout, because another cat had 'written its name' on me. To Nip, that meant that I no longer 'belonged' to her, but to another cat, which had 'marked' me!

Pheromones are produced by **exocrine** glands. [<u>Ex</u>-, as in 'exit' and

Science 2 *for Little Folks*

'external,' means *outside*.] Exocrine glands send things like pheromones and sweat *outside* the body. [Have you ever seen beads of sweat on your forehead or nose? Your sweat glands are exocrine glands.] Other glands send chemical messages inside the body, but glands that produce pheromones send invisible messages *outside* the body to others.

Not only cats, but other animals have pheromones. Mice have pheromones that, smelled by other mice, warn of danger. Deer mark their territory with pheromones. Humans can't smell most pheromones, but animals and insects that need to 'get the message' smell them just fine.

Insects use pheromones to send many different kinds of messages. Ants leave trails marked with pheromones to tell other ants the way to go to find a piece of food. When the food is all gone and the ants stop marking the trail, the pheromone trail fades away and the ants stop using it. Ants use another pheromone to send a "Help! Danger! Come quickly!" message. When other ants smell the message, they hurry to help defend the ant colony.

Butterflies use pheromones to locate other butterflies that are too far away to see. If your mom was a few miles away, you couldn't see where she was, and if neither of you had a telephone, how would you find each other? For butterflies, it is easy. They can smell the pheromones of, and thereby find, another butterfly as far as five miles away.

Did you know that God always knows where *you* are, too? That's because He is always with you! No matter how far away, or how dark the night, God is always there. He has also marked you with an invisible sign. When you were baptized, God marked your soul with a beautiful spiritual mark called a *character*, which lasts forever. Unlike the ants' trail, which will fade away, you have been marked with a sign of Christ which can never be removed. You belong to Jesus.

"You...are Christ's sheep; you bear the mark of the Lord in the sacrament you have received..."

—St. Augustine: Letter 173

*"...from Your presence where can I flee?
   if I settle at the farthest limits of the sea,
even there Your hand shall guide me
   and Your right hand hold me fast.
If I say, 'Surely the darkness shall hide me,
   and night shall be my light'—
for You, darkness itself is not dark
   and night shines as the day."*
   —Ps. 139:7, 9-12 NAB

Questions

1. Animals and insects use this to mark territory and send messages.

   pheromones

2. What kind of glands send things like pheromones and sweat outside the body?

   exocrine glands

3. Name an exocrine gland that you have!

   sweat gland

4. At Baptism, God marks your soul with an invisible mark called a:

   Character

Zoology

How to Dress a Duck

Have you ever gone wading in a creek or pond in the wintertime? Brrr! The water is so cold, it almost makes your teeth rattle, doesn't it? But if you have ever seen a flock of ducks paddling about in that same frigid water, they seem quite happy. How can they be so comfortable in such COLD water? Do you suppose that God knows how to 'dress a duck'? Let's find out!

If you want to stay toasty warm, what is the first article of clothing that you put on? Why, long underwear, of course! And that is what God did with ducks, too. He began with soft, cozy 'long underwear' called *down*. Down is a fluffy layer of fine feathers with tiny air pockets, to keep warmth right next to the duck's skin.

But what good does long underwear do if it gets all wet and soggy? God has planned for that, too. Over the down, God placed a shiny 'raincoat' of outer feathers, smooth and well-oiled. *Preening* spreads oil from a large gland next to the tail. During preening the duck uses his bill to take oil from the gland and spread it carefully up and down each outer feather. The oil from preening waterproofs the feathers, which are very closely fitted together to help keep the water out, also.

What else would a duck need to 'wear' while swimming? How about a life jacket, to keep him safely afloat? God thought of that, too. The duck's feathers are hollow and filled with air; between the trapped air in the down and the air in the hollow feathers, Mr. Duck has a lovely 'life jacket' that pops him right to the surface of the pond.

Let's see—Mr. Duck has cozy long underwear, a shiny raincoat, a lifejacket and...SWIM FINS! Have you ever used swim fins? [You wouldn't want to use them unless you are already a good swimmer, like the duck.] But if you can swim well, a pair of swim fins can make you ZOOM through the water. Mr. Duck has webbed feet; like swim fins, webbed feet have no separate toes, but instead are shaped a lot like a wide paddle or fan. With his neatly designed feet, he can swim quickly underwater to find his food or make a fast getaway from other animals who might want to eat him for dinner!

Finally, we come to Mrs. Duck; she has all the neat 'clothes' that Mr. Duck has, but God thought of one more thing just for her. Mr. Duck usually has brightly colored feathers; he is a pretty fancy-looking fellow. But God planned something special for Mrs. Duck, and special 'clothes' for the job. Mrs. Duck sits for hours at a time on her nest of eggs, waiting for her babies to hatch. Her nest is generally in the dried, brown grass next to the pond. If she had bright feathers, she and her nest would be easy to see. An animal might find her and eat both Mrs. Duck and her eggs! However, God made Mrs. Duck's feathers a lovely brown to blend right in with the grass. Her color helps protect her and her babies.

God knows everything needed to 'dress a duck,' for sure. How much more does Our Lord know how to take care of all the needs of His children.

"...God will supply every need of yours according to His riches in glory in Christ Jesus."

—Phil. 4:19

Questions

1. A fluffy layer of fine under-feathers is called what?

   _____

   _____

2. What does the duck do to spread oil on its feathers?

_____

_____

3. Describe webbed feet.

_____

_____

_____

_____

Experiment!

Have you ever heard the expression, 'like water off a duck's back'? You can see for yourself what it means in this experiment. You can also see how oil makes a duck's feathers waterproof.

Cut one side out of a paper grocery sack. Working over the kitchen sink, pour one teaspoon of vegetable oil onto the center of the cut-out piece. Smooth the oil into a circle, about the size of your hand, until it is all absorbed by the paper. Put the paper in the bottom of the sink. Now run a trickle of cold water over the paper. Does the water behave differently when it flows over the oiled part of the paper than when it flows over the un-oiled part? Now tilt the paper a little. The water will make little 'beads' on the oily part.

Can you see that the water doesn't 'stick' to the oiled part of the paper, but runs 'like water off a duck's back'? Because oil sheds water, it is also often used to protect machines from damage caused by water. Ask a parent to show you some of the machines around your home that are oiled for protection.

Biology

Circulatory system: Blood

Precious Blood

While you play and as you sleep, life-giving blood flows through little 'hoses' in your body. Some of these 'hoses,' called *veins* and *arteries,* can be seen under your skin.

Look at your wrists. Can you see the blue veins? Sometimes, if you take a strong-beamed flashlight into a dark room and put your palm over the light, you can see the blood vessels in your hand. The red glow that you see is caused by the blood flowing through vessels as thin as threads. These tiny vessels are called **capillaries**.

Through these blood vessels flow all the cells that together make up your blood. The *red cells* carry through your body the oxygen from the air which you breathe. A person must have oxygen to live. If his blood suddenly stopped carrying oxygen, he would be dead within seven or eight minutes.

As important as the red cells are, **platelets** and *plasma* are the parts of the blood that make your blood *clot*. A clot is a special kind of 'plug' made of plasma and sticky platelets. When you get cut, the clot plugs up the hole in the cut blood vessel and makes it so no more blood can come out of the hole. Then the clot turns into a scab. If your blood did not clot, you could bleed to death from just a small cut.

Blood also contains *white cells*. White cells are 'fighters' that travel through

Science 2 *for Little Folks* **7**

your blood stream looking for germs. When they find germs, they attack! Have you ever had a cut or a sliver that became infected? If you have, you probably remember seeing some yellow 'stuff' called *pus* around the injury. That pus was actually white cells at work ridding your body of germs.

All the different parts of our blood work to keep us strong and healthy, by carrying oxygen and food through our bodies and getting rid of waste and germs that might make us sick.

Can you see what a precious gift Our Lord gave you? *Precious* means 'of great value.' Our blood is of great value to us, isn't it? But there is a Blood that is far more precious than ours. That is the Precious Blood of Our Lord, which He poured out for us on the Cross. His Precious Blood makes us spiritually strong. It gives us the grace to fight off sin, and heals our injured souls. Jesus' Precious Blood is a gift of great value, *infinite*, everlasting value, that makes it so we can live forever in Heaven with Him. Our blood keeps us alive here on earth, but it is the Precious Blood of Our Lord that gives us eternal life.

"He who eats My Flesh and drinks My Blood abides in Me and I in him."
—Jn. 6:56

"Let us fix our gaze on the Blood of Christ and realize how precious it is to the Father, seeing that it was poured out for our salvation and brought the grace of conversion to the whole world."
—Pope St. Clement I [1st century]

Questions

1. 'Hoses' that carry blood are called:

   _____

   _____

   _____

2. Which blood cells carry oxygen?

   _____

   _____

3. Platelets and plasma make the blood do what?

   _____

   _____

4. Which blood cells fight germs?

   _____

   _____

5. Find *precious* in a dictionary and write the definition.

   _____

   _____

   _____

   _____

Science 2 for Little Folks

Biology

Skin: Pigmentation

Coloring Adam and Eve

What color was Adam and Eve's skin? What color is your skin? Is it black, pink, yellow, red, brown, or white? Would you be surprised to learn that it is brown?

Do this test: fold a piece of white paper in half. Unfold the paper and, using the fold line as the bottom line, color black, pink, yellow, red, and brown squares about an inch wide right at the fold. Now fold the paper at the line again, so the squares are right at the edge of the paper. Hold the paper against the back of your hand. Do any of the colors, including the white of the paper, match your skin color?

You probably didn't find any color crayon that exactly matched your skin color. Let's find out why [unless you are an albino] your skin is actually some shade of brown.

In your skin is a *pigment,* or coloring, called **melanin**. Melanin is a brown pigment, made in your skin by cells called **melanocytes**. No matter what their 'race,' all human beings have about the same number of melanocytes and all have the brown pigment melanin! However, people with lovely dark skin produce more melanin that those with lighter skin.

It is a truth of our Holy Faith that we all descend from our first parents,

Adam and Eve. It is fortunate that we inherited from them the ability to produce melanin, for it helps protect against the harmful effects of too much sun. Melanin, in both the skin and eyes, absorbs or 'soaks up' some of the sun's rays, giving some protection from skin cancer.

Being out in the sun makes your skin produce more melanin, which makes your skin a darker brown. It is melanin that gives you a tan. And freckles are 'spots' of melanin.

If you have read 'Eyes to See,' you know about the iris, the colored part of your eye. Jesus put melanin in your eyes, too, to help protect them from the glare of the sun. If you have lots of melanin in your eyes, they are probably a yummy chocolate brown. If you have only a little melanin, the iris is probably a perky blue. But whether your eyes are blue, brown, hazel or green, it is melanin that gives their color.

Not only does melanin color your skin and eyes, but your hair, too. The more melanin in the hair, the darker it is. As people age, the hair often gradually loses its ability to make the pigment, so the hair turns white. People whose bodies produce no melanin at all have white hair, even when they are children.

A person whose body produces no melanin is called an **albino**. Albinos have to be very careful about staying out of the sun. With no melanin for protection, both skin and eyes are very sensitive to sunlight. Sun burns and skin cancer are very serious problems for people who have no melanin. Can you see that it is a good thing to have lots of melanin?

How fortunate we are to have the protective 'brown shield' of melanin, that we inherited from our first parents. But did you know that we also have a protective 'blue shield'? It is the mantle of our Mother Mary!

Mary, Help of Christians, pray for us!

"She, by the very fact that she brought forth the Redeemer of the human race, is also in a manner the most tender mother of us all, whom Christ our Lord deigned to have as His brothers [Rom. 8:29]."
—Pope Pius XI

Questions

1. Skin's brown pigment is called:

_____

_____

2. Melanin is made by cells called what?

_____

_____

3. People whose bodies produce no melanin at all have what color hair?

_____

_____

4. With no melanin for protection, both skin and eyes are very sensitive to what?

_____

_____

Biology

Exocrine glands: Lacrimal glands

He Will Dry Every Tear...

When you cry, do you end up with a pile of soggy tissues because both your eyes and your nose are 'running over'? Why does your nose run when you cry? Let's find out!

Your tears pour from the **lacrimal glands**, also called tear glands, located under the upper eyelid on the outer side of your eye. [The lacrimal glands, which empty their contents outside the body, are part of the system of exocrine glands. You can read more about exocrine glands in the story, 'Invisible Marks.'] Since the glands are located in the eye, it seems odd that they can make your nose run, doesn't it?

Because your eyes could be damaged if they weren't kept moist, God designed the lacrimal glands to produce just the right amount of tears to keep the entire surface of the eye constantly bathed in moisture. He also made a *tear duct,* which is a tube-like kind of drain to empty away extra fluid from the tear glands. Had God placed the tear duct directly under the lacrimal gland, the tears would just have run down into the drain without first covering the whole eye. Then part of the eye would have been dry and unprotected. But our wise Designer put the tear duct at the opposite edge of the eye, right next to the nose! Placing the gland on one side

Science 2 *for Little Folks*

and the duct on the other causes the tears to pass all the way across the eye before draining into the nose. When the tears drain into your nose, they make your nose run!

Besides keeping your eyes moist, tears serve to clean the eye of little pieces of dirt or sand. When a 'foreign' particle gets in your eye, it makes the tear glands produce a 'flood,' which helps wash out the irritating particle. The shape of the eye and the blinking of the eyelid helps distribute the protective tears. In addition, tears contain salt and other substances that help protect the eye from bacteria, or germs, and infection.

Does it seem to you that women cry more than men? They do, and for a good reason. Scientists have discovered that the same chemical which causes a woman's body to make milk for her babies also helps produce tears. God put this chemical in both men's and women's bodies, but women have about twice as much as men do. So women usually make more tears!

Scientists also think that crying when we are sad is probably good for our health!

St. Monica cried from worry about St. Augustine's sinful behavior before he gave his life to Our Lord. In St. Augustine's *Confessions*, he writes about his mother visiting a bishop to talk with him about her son. The bishop told her not to worry, because God would honor her prayers and St. Augustine would come back to God. "Go your way; as sure as you live, it is impossible that the son of these tears should perish."

Do you suppose that our Blessed Mother ever cried? The Bible doesn't tell us for sure, but one of the Blessed Mother's special titles is 'Our Lady of Sorrows.' With her most tender of hearts, surely she wept, as did Our Lord, when she saw the suffering caused by sin. How deeply Mary must have suffered at the foot of the Cross. Yet, how Our Lady's eyes must have filled with tears of joy at her Son's Resurrection and Ascension!

The next time you feel the tears start to roll down your cheeks, think of Our Lord on the cross, with Our Sorrowful Mother at His dear feet. Then thank them for the tears and suffering which they offered for your salvation.

"'Standing by the cross of Jesus was His mother' [Jn. 19:25]. The virgin, with her mother's grief, participated in a quite particular way in the Passion of Jesus..."
—Pope John Paul II

"...and God will wipe away every tear from their eyes."
—Rev. 7:17

Questions

1. What do we call the glands that produce tears?

   _____

   _____

2. What is the name of the tube-like kind of drain into which the tears drain?

   _____

   _____

3. What do tears contain that help protect the eye from bacteria?

   _____

   _____

Something to Do

1. Ask your Dad or Mom to help you find out more about the Seven Sorrows of Our Lady.

2. Read the 11th chapter of the Gospel of John for an account of Jesus' tears.

Science 2 *for Little Folks*

Biology

Circulatory system: Heart

Our Hearts and Theirs

Have you ever run or played so hard that you thought your heart was going to pound right out of your chest? I'll bet that you were panting for breath, too. When we exercise *vigorously,* our bodies need more food and oxygen for our hard-working muscles. It is our muscular hearts that help deliver that food and oxygen at just the right rate to meet the needs of the moment.

Normally, when you are resting, your heart muscle beats so quietly that you probably don't feel it at all. It just gently does its job of pumping the blood that carries the *nutrients* throughout your body. [Nutrients from digested food are what give you energy and help you to grow properly.] When you are resting, the heart beats at a rate of about 90 beats per minute, but when you exercise, your heart speeds up. The speed at which your heart beats is called the *heart rate.*

However, heart rates *vary.* In brand-new babies, the heart rate may be as high as 130 beats per minute; an adult has a heart rate of about 72. Usually, the smaller the person [or animal], the faster the heart beats. Because of this, women generally have higher heart rates than men. [The tiny hummingbird has a heart rate of about 800 beats per minute; the heart of the massive elephant beats only 25 times per minute!]

Whether human or animal, the more work the body does, the more it needs nutrients and oxygen. Then the heart pumps faster to provide the food and oxygen that is needed. The heart faithfully pushes the blood through the *arteries,* which are the blood vessels carrying blood from the heart to the rest of the body. Without a heart, we could not live.

Perhaps it is because the heart is such a faithful servant, or perhaps because the heart is absolutely necessary to life, that when we think of hearts, we often think of love. If we say that we love someone 'with all our heart,' it is another way of saying that we love them with our whole being, that we would give our lives completely to them. The Blessed Virgin Mary loved Jesus with her whole Immaculate Heart. Now, when something is 'whole,' it is also *pure.* That is, if you have a piece of pure gold, it has nothing else in it but gold; it is wholly, purely gold. So Mary's *pure* [Immaculate] heart's *entire* [whole] purpose was nothing else but love of God. She so united her Heart with His that she lived her whole life completely and perfectly joined to His will. Our Lord, with His Sacred Heart, loved us so much that He gave His life completely for us, even to having His Sacred Heart pierced to pour out every last drop of His Precious Blood on the Cross. His pierced Heart opened the way, so that when our hearts finally do stop beating, we might gloriously join Him for all eternity in Heaven. May our hearts be united with Theirs forever.

"Let us learn to cast our hearts into God."
—St. Bernard

"Even as our predecessor of immortal memory, Leo XIII, ...saw fit to consecrate the whole human race to the Most Sacred Heart of Jesus, so we have likewise, as the representative of the whole human family which He redeemed, desired to dedicate it in turn to the Immaculate Heart of the Virgin Mary."
—Pope Pius XII

Questions

1. Using a dictionary, define *vigorously* and *vary.*

   _____

   _____

   _____

2. What are the blood vessels that carry blood from the heart to the rest of the body called?

_____

_____

3. What is the heart rate of a newborn baby?

_____

_____

4. What is the heart rate of an elephant?

_____

_____

Experiment!

The *pulse* is the throbbing caused by the heart pushing blood through the arteries. By taking your pulse, you can determine your heart rate. Find the pulse in your **carotid artery** by gently placing your middle two fingers flat across the side of your neck, with the ends of your fingers just touching under the angle of your jaw. Using a watch with a second hand, count how many heartbeats you feel for one minute. Write down the number. Then run in place or jump rope for two minutes. Now take your pulse again. Was your pulse higher after exercising?

Biology

Exocrine glands: Hormones

Run for Your Life!

Have you ever been startled by a snake slithering through the grass, or a growling dog, and suddenly found that you could run really, really fast? A grandma whom I know once saw a grass fire burning toward her house and raced outside to put it out. Twice, to bring more water to throw on the fire, she jumped right over a five-foot tall wire fence! Where did her extra strength come from?

You may have read about **exocrine** glands in the story 'Invisible Marks.' Exocrine glands send chemical messages outside the body. Our Lord also designed a system of **endocrine** glands to send chemical messages *inside* the body. He designed these glands to pour their *hormones*, or chemical messengers, directly into the bloodstream so they could begin working *fast*. That is why you can move so quickly when you hear a growling dog behind you!

The endocrine glands that give you quick, 'run-for-your-life' energy are called the **adrenal** glands. In an emergency, the adrenals send out the hormones **epinephrine** and **norepinephrine**. These chemical messengers tell your heart to beat faster and cause more blood to flow to your brain and muscles, so you can think and respond quickly. The adrenal hormones even make your pupils grow in size to let in more light. Then you are better able to see what startled you! God even

Science 2 *for Little Folks*

designed the adrenal hormones to make your blood clot and form scabs faster, in case you run so fast that you fall and skin your knee.

Have you ever been a little nervous when it was time to do something important and exciting? Perhaps you were excited about meeting our merciful Lord face to face the very first time you went to Confession. [You don't need to be nervous, you know. He is just waiting to wrap you in His loving arms.] Maybe you were a shepherd or an angel in a Christmas play. Afterwards, did it take you a little while to calm down? Your adrenal glands were at work! Sometimes, if the exciting event happens late in the evening, it is hard to fall asleep at bedtime, isn't it? That is because God made the adrenal hormones to keep you awake and alert, ready to meet any challenge.

The endocrine system includes other glands besides the adrenals. One of these is the **pituitary** gland, which is only the size of a small marble. The pituitary gland sits just under your brain. It sends out several different hormones and helps tell some of the other glands what to do.

By means of chemical messengers, the hormones, Our Lord gives us strength and energy to deal with scary situations. Through His angelic messengers and His own mighty power, He also strengthens us to flee temptation and run safely toward our heavenly goal.

"God is our refuge and our strength, an ever-present help in distress. Therefore we fear not, though the earth be shaken and mountains plunge into the depths of the sea."

—Ps. 46:2-3 [NAB]

*Soldiers of Christ! arise
And put your armor on,
Strong in the strength which God
   supplies
Through His eternal Son;
Strong is the Lord of Hosts,
and in His mighty power,
Who in the strength of Jesus trusts,
is more than conqueror.*

—ANONYMOUS

Questions

1. ___ glands send chemical messages inside the body.

   _____

   _____

2. Another name for *chemical messengers* is:

   _____

   _____

3. Name the hormones that are produced in the adrenal glands.

   _____

   _____

4. Which endocrine gland sits just under the brain?

   _____

   _____

Biology
Eyes

Eyes to See

Driving home late at night, have your car's headlights ever caught the 'glow' of an animal's eyes, shining back at you through the darkness?

Cats, deer, sheep, and other animals that feed or hunt at night have a very special 'something,' called **tapetum lucidum** in their eyes. Tapetum lucidum works a little like a bicycle reflector, causing the animal's eyes to bounce light back at you; that is why you can see the eyes 'glow' in the dark. More importantly, tapetum lucidum also causes light to shine back into the eye of the animal. This makes it so the animal can see better in the darkness.

Do your eyes 'glow' in the dark? No, because God did not create you for the same purpose as a deer or a cat. He knew that you would not be prowling at night, hunting tasty mice for your dinner, so He did not put tapetum lucidum in your eyes.

God did, however, make a way for your eyes to let in more, or less, light. The colored part of your eye is called the *iris*. [Do you have pretty chocolate-brown iris, or perhaps perky green or hazel, or maybe even pale blue?] At the center of your colorful iris is the *pupil*. The black pupil is really just a hole in the center of the iris. When the eye needs more light to see, your nerves tell the muscles of the iris to open up wider and let in more light. This makes the pupil bigger.

On the other hand, if the light is too bright, the iris squeezes part way shut to shrink the pupil and protect the eyes. [If the light is too bright, it can injure your eyes. You should never look right at the sun, for example, for that can damage your eyes. But, in addition to causing your pupils to shut out most of the light, God gave you an extra 'safety switch.' If you should accidentally look at the sun too long, God made your nerves to react so you would shut your eyes completely and move your head, too, by making you sneeze!]

Do you ever think about things that you can't see? Can you see God's grace, which is His life in us? No, it is invisible. But He gives us the sacraments so we can 'see' His grace through them. [Sacraments are an outward sign of God's grace, which cause what they 'sign.'] In His wonderful gift of the Sacrament of Baptism, we can see the flowing waters by which God washes away sin and gives life to the soul, through the actions of the priest. In the sacraments, God pours His life into us, using ways which we can hear, taste, touch, feel, and see.

Thanks be to God that He gave us eyes to see the beautiful world around us. Let us also use them to watch closely as He becomes present to us at Holy Mass and in all the sacraments.

"Blessed are the eyes which see what you see! For I tell you that many prophets and kings desired to see what you see, and did not see it..."
— LUKE 10:23

Questions

1. What is the colored part of the eye called?

   _____

   _____

2. What is the black center part of the eye called?

   _____

   _____

3. Is it a good idea to stare at the sun? Why or why not?

_____

_____

_____

_____

_____

_____

Experiment!

Would you like to see the muscles of the iris and the pupil at work? Take a flashlight and go into the bathroom, but don't turn on the light. Count to 300, then turn the flashlight on and quickly shine it into the mirror. Watch your pupils carefully. Were they very big at first? Did they shrink while you watched? The pupils widened in the dark to let in more light, but got smaller when the bright light of the flashlight shone on them.

Biology

Nervous system: Nerves

Nerves and Your Sense of Touch

"**Ouch! Hot!**"

Is that what you said the last time you burned yourself touching a hot pan or another hot surface? Did you hurry to the sink and stick your hand under cold running water? I hope so, because that is a good way to help lessen the burn and the pain, too.

How did you know that the pan was hot? Why, it was your *nerves* that told you! Your nerves also told you to move your hand, *fast!* **Receptor** nerves, [or **neurons**, which is what nerve cells are called], are designed to sense hot and cold and touch. Many types of neurons run, almost like little telephone wires, from your fingers right up your arms and through the *spinal cord*. [The spinal cord is a long bundle of nerves that runs up the center of your back bone.] Usually, the spinal cord carries the messages to your brain, which is your 'command center.' But if you touch something hot, God has designed the receptor neurons to 'take a short cut' so the message to 'MOVE' gets through even faster. The message is picked up right in the spinal cord and sent to *motor neurons*, which scream the message to your arm and hand muscles: "Move it!"

What might happen if your receptor neurons didn't work? What would happen if you picked up a hot pan, but didn't know that it was hot? You might continue holding it and burn yourself

Science 2 for Little Folks

so badly that you might forever lose the use of your hand! Can you see how dangerous it might be if your neurons were damaged and you were not able to 'feel'?

The Bible mentions a disease which damages the neurons. The disease is called *leprosy*. A person with leprosy loses the ability to sense hot and cold and pain. Partly as a result of this nerve damage, lepers in Jesus' time often were missing fingers and toes and other parts of their bodies. A beautiful passage from the Gospel of Luke tells how Jesus showed His great love by healing ten men suffering from leprosy. Sadly, only one of the men remembered to thank Him. "Then one of them, when he saw that he was healed, turned back praising God with a loud voice; and he fell on his face at Jesus' feet, giving Him thanks." [Lk. 17:15-16]

How fortunate we are that Our Lord designed our nerves and nervous system for our protection. The next time your neurons tell you to *Move!*, remember to 'praise God with a loud voice!'

Questions

1. What is another name for nerves?

   _____

   _____

2. What is the name for the long bundle of nerves that runs up the center of your back bone?

   _____

   _____

3. What is the name of a disease that damages the neurons' ability to feel hot and cold and pain?

   _____

   _____

More About Nerves

Besides having nerves in your hands and arms, you have them all through your body. The nerves do other jobs besides 'feeling.' For example, some neurons help you learn about the world around you through your sense of smell, taste, vision, and hearing. Even when you are not aware of it, your 'command center,' the brain, sends and receives messages through the nerves. Some special neurons tell your heart to beat and your lungs to breathe, even while you sleep! All of your nerves together make up what is called your *nervous system*.

Experiment!

With a toothpick, poke one hole at the top of a piece of notebook paper. Then poke two holes in the center of the paper, about 1/8" apart. At the bottom of the paper, poke three holes, also about 1/8" apart. Now turn the paper over, so the 'bumps' that the holes made stick up. With your eyes closed, can you feel which group has one, two, or three bumps? God filled your fingertips with *sensory* nerves so that your fingers are very sensitive to small differences in texture and 'touch.'

Here is a trick to try on Mom or Dad. Make a 'V' with your first and second fingers outstretched. Say, "Close your eyes." Now, touch those two fingertips to the middle of Mom or Dad's calf. [The calf is the muscled part of your lower leg, halfway between your knee and your ankle.] Ask, "How many fingers do you feel?"

Switch back and forth between one finger and two, asking the same question each time. Is one part of the leg more sensitive than another? Can your parents tell how many fingers you used? Now, put the paper with the bumps on it against the calf. Can Mom or Dad tell how many bumps are on the paper? Can they feel the bumps at all? The leg isn't as 'sensitive' as the fingertips, is it? God gave your legs a different 'job' than your fingers, so He gave the legs fewer sensory nerves.

Science 2 *for Little Folks*

Biology
Hearing

Ears to Hear

Do you ever sit outside and, with closed eyes, thank Jesus for His awesome gifts, including the gift of hearing? I like to keep my eyes closed and try to identify all the sounds that drift up from the creek below our home. Today, the sounds of leaves rustling in a soft breeze, insistent 'chirps' of a chipmunk, the throaty purr of the cat at my feet, and water burbling as it spilled over a beaver's dam, all came to me through the gift of hearing.

How do we experience sound? Sounds travel as *vibrations*. You can feel vibrations by placing your palms flat on a table while someone 'drums' the table-top with his fists. Even if you are not touching the table, you can hear the vibrations in the air. The vibration from the pounding has also caused the air to vibrate. These vibrations of the air are called *sound waves*.

Try this: fill a tub with four or five inches of cold water. Let it sit until the water is perfectly still. Then very gently strike the surface of the water. Can you see little ripples, or waves, moving from your hand to the edge of the tub? When you touched the water, it 'pushed' the water, which pushed the water next to it, which pushed the water next to it, making waves.

In a similar fashion, vibrations from sound strike the air, causing it to strike the air next to it, which strikes the air next to it, making sound waves. Unlike the waves in the water, sound waves cannot be seen. But when the sound waves reach your ear, you can hear them!

Very high sounds, or sounds with a high *pitch*, are made by faster vibrations. Low-pitched sounds have slower vibrations. Usually, you can feel the vibrations of lower-pitched sounds more easily than those that are of high pitch. However, some very low-pitched sounds are just as hard to feel and hear as extremely high-pitched sounds! Feeling the **larynx**, also called the voice box, can show you the difference between low-pitched and high-pitched vibrations. Put your fingers on your larynx, and hum first low tones and then high tones. Which tones make your larynx vibrate more?

Our Lord created human ears to hear a different range of sounds than animals do. For example, bats are able to hear extremely high sounds that people cannot hear. God gave them this ability for a good reason. Many bats eat insects, that also make high-pitched noises. Now, bats can't see very well, and they are mostly out and about looking for food at night. So God gave bats a special way not only to hear the insects, but to bounce sound waves off the insects, and other objects, as a way to help find them in the dark!

On the other hand, whales and elephants can *communicate*, or 'talk,' over a distance of several miles using extremely low-pitched sounds. These sounds, too, are far beyond the limits of human hearing.

Now and then, it is not the pitch of the talking, but whether or not we want to hear what is being said, that makes it 'easier' to hear. When you are busy playing, have you ever noticed that it is easier to hear "Ice cream!" than "Please pick up your jacket."? Our ears need to be trained to listen to all that God calls us to do. In fact, the most important listening we do doesn't use our ears at all! The voice of God is best received by well-trained 'ears of the heart.'

How lovingly Our Lord's tender voice calls to our 'spiritual ears'! He speaks to us through His Holy Church, His Word, His Blessed Mother and His holy priests. Listen:

"Fear not... I have called you by name, you are Mine...you are precious in My eyes..."
—Is. 43:1 & 4

"Listen...let nothing discourage you...are you not in the folds of my mantle and the crossing of my arms?"
—Our Lady of Guadalupe

How sweet the sound!

"...not everyone hears clearly. All ask what they wish, but do not always hear the answer they wish. Your best servant is he who is intent not so much on hearing his petition answered, as rather on willing whatever he hears from You."
—St. Augustine

"Do you not yet perceive or understand? ...having ears do you not hear?"
—Mk. 8:17 & 18

Questions

1. Sounds travel as:

_____

_____

2. Vibrations of the air are called:

_____

_____

3. Another name for the voice box is:

_____

_____

Something to Do

Put your favorite music for praising God in the tape or CD player. Now blow up a balloon and hold it very gently between your palms. Hold it about three or four inches in front of the speaker as the music plays. Can you feel the vibrating air inside the balloon? The vibrations are caused by sound waves from the music.

Biology

Digestive system: Saliva

Spit!

This morning I awoke to the sounds of a raccoon fight in the back yard. Two young raccoons were apparently having a disagreement over breakfast and which of them was going to eat it! Perhaps a juicy apple was the center of the argument, or a tender frog caught in the bubbling creek just down the hill.

Maybe you have seen photos of raccoons washing their food before eating it. People used to think that raccoons didn't have any **salivary glands,** and that was the reason they so often washed their food. [Salivary glands make *saliva,* which children sometimes call 'spit.'] We now know that raccoons do have salivary glands and saliva, and it's a good thing, too, because their little mouths would become awfully dry without it!

Saliva moistens both your mouth and your food, which helps make it so you can swallow what you eat. [Have you ever tried to eat two soda crackers at once? It takes a lot of chewing to 'get them down,' because they absorb most of your saliva. As you continue chewing, your salivary glands produce more saliva, which softens and moistens the crackers.] How could you eat if you did not have saliva?

After you have swallowed your food, it goes to the stomach to be digested. But digestion actually began in your mouth! Saliva contains something that begins the digestive process, which then continues in your stomach.

While starting the digestive process and moistening your mouth, your 'spit' also helps clean your teeth. [Of course, you must help by brushing your teeth carefully after you eat!] Some people have very little saliva and have to clean their teeth extra well because, without the help of saliva, they can develop really bad cavities.

Do you ever eat in bed? I hope not! God knew that you could not eat and sleep at the same time. Since you would not need saliva to moisten your food then, He designed your salivary glands to produce very little saliva while you sleep. So it is important to go to bed with clean teeth, because your salivary glands are 'taking a rest' from cleaning your teeth, too!

One way that saliva may help fight cavities is that it contains something that fights germs. Have you ever seen a dog or a cat licking a sore spot? A substance in the saliva helps the wound to heal. People have germ-killing 'stuff' in their saliva, also. Now, your saliva also contains germs, so you don't want to run around licking everybody thinking that you are doing them a favor. You just might make them sick! But you can see that, in creating something as simple as saliva, Our Lord was planning every detail for our good health.

"...and plaiting a crown of thorns they put it on His head, and put a reed in His right hand. And kneeling before Him they mocked Him, saying, 'Hail, King of the Jews!' And they spat upon Him...."
—MATT. 27:29-30

"As He passed by He saw a man blind from his birth... He spat on the ground and made clay of the spittle and anointed the man's eyes with the clay, saying to him, 'Go, wash in the pool of Siloam.' So he went and washed and came back seeing."
—JN. 9:1; 6-7

Questions

1. What is another word for 'spit'?

   _____

   _____

2. Saliva is made in what glands?

   _____

   _____

3. Where does the digestive process begin?

   _____

   _____

4. Why is it important to go to bed with clean teeth?

   _____

   _____

Think and Pray

Everything that Our Lord has created, even saliva, is good and meant to be used for good. God gave us a free will, that we may choose to do good with the things that He has made. Sadly, sometimes people choose to do wrong with the good things that God created.

Take two or three minutes to sit or kneel down before an image of Our Lord on the cross and read the two verses from this story. Think about the two very different ways that saliva was used in these verses. Can you see how Our Lord can either be mocked or glorified even with simple things in our lives?

Let us choose to glorify our loving God and Savior, Who has given us every good thing.

Science 2 *for Little Folks*

Health

Nutrition

Soul Food

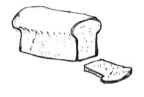

Do you sometimes have peanut butter and banana sandwiches for lunch at your house? I like peanut butter and banana sandwiches because they are delicious. I also like them because they are *nutritious,* or full of good things that bodies need to be healthy. These sandwiches are full of *protein*, to build strong bodies. They also have *fat* and *carbohydrates,* to give you energy to think and jump and run.

To grow properly, your body needs protein, carbohydrates, and a small amount of fat every day. But that doesn't mean you have to eat only peanut butter sandwiches! You can get *protein* from meats, fish, and foods made from milk. *Fats* come both from animals and plants. Meats and dairy foods like milk, cheese, butter, and ice cream all contain some fat. Beans have a small amount of fat, but nuts have a lot! *Carbohydrates* are found mostly in grains and legumes. [Grains are foods like wheat, corn, oats, and rice. Some legumes are beans, peas, and peanuts.] You can see that eating many different kinds of food each day will help you get all the nutrients that you need.

Now you know that you can look in the meat section of the grocery store to find protein foods. But wait! You can also get protein by eating *grains* and *legumes* together. That is one reason why peanut butter sandwiches are so healthful, because they are made with bread from grains, and peanuts, a legume. The peanuts and grains in your sandwich give you protein, fat, and carbohydrate, all in one package. And you didn't even have to stop in the meat department to pick up your protein. So you are all set for meatless Fridays!

Let's think for a moment about what you eat most. I'll bet you have some type of grain at nearly every meal. Was cereal on your breakfast table this morning? Since *cereal* is another name for grain, if you had toasted bread, you also had cereal. Breads and pasta are made of grains, too. Of all the different foods that people eat around the world, grains are eaten most of all.

In most countries, wheat and rice are the favorite grains. However, corn and oats are popular, too. Most of these cereals are made into some kind of bread that is eaten at almost every meal. You can understand why bread is sometimes called 'the staff of life.' That is another way of saying that bread 'holds us up,' or 'bread is where we find our strength.'

Then there is another type of Bread that 'holds us up' and gives us strength. This Bread, however, is not food for our bodies, but food for our souls. It begins as a bread 'which earth has given, and human hands have made.' The Church teaches that it is to be made only from wheat and water, with nothing else added. Can you guess what Bread this is?

It is the Bread of Life, our Jesus, our Eucharistic Lord! Of course, you know that at Holy Communion, it is bread no longer; it only looks like bread. At the Consecration, it has become Our Lord. This is a great Mystery of our Faith.

Sometimes people have a hard time believing this Mystery, but I will tell you another Mystery that is just as true: on the first Christmas, Almighty God, Who loves us so much, came to earth to live with us! No, He did not come in all His power and might. If He had, everyone would probably have been scared to death. Instead, He became a tiny Baby, the Infant Son of the Virgin Mary, but He was Almighty God at the same time.

Now, think about this: as tiny as He was when He was born, Jesus was still God. In fact, He was God the day before He was born, and the week before that, and the month before that. Just the same way that you were still *you* on the day that you were born. Sure, you were smaller then, but you were still you—even while you were inside your mother. The whole time that Jesus was growing under His mama's heart, He was God, right back to the day that His life began. [You know about the Annunciation, don't you?]

So God, Who made everything that is, Who is bigger than the whole world, bigger even than the immeasurable universe, made Himself very, very small for love of us. Of course we know that this great and wonderful Mystery is true. But there were some people, here and there, who saw Jesus while He was living on earth, who did not 'see' that He was God. Perhaps they were not present at His mighty miracles; maybe they did not hear His words of life. For whatever reason, some saw Jesus as only a man.

And what does this have to do with bread, or Bread? Well, some people have never learned the whole Truth about Jesus' Real Presence at Holy Communion. And the Truth is this: the God of all creation chose to become a tiny Baby so that He could be with the people He loves so dearly. In much the same way, that same God has also chosen to make Himself so small that He can come to live inside each one of us as the Bread of Life.

"'For the bread of God is that which comes down from heaven, and gives life to the world.' They said to him, 'Lord, give us this bread always.'

'Jesus said to them, 'I am the bread of life...'"

—JN. 6:33-35

Questions

1. *Nutritious* means 'full of good things that bodies need to be....' what?

   _____

   _____

2. *Protein* builds strong what?

   _____

   _____

3. What two things give you energy to think and jump and run?

   _____

   _____

4. An easy way to get *carbohydrates* is to eat what kinds of foods? Name at least two.

   _____

   _____

Biology
Viruses and bacteria

Germs Make Me Sick!

"But, Mom, I don't see any germs!" Did you ever say something like that when your mother told you to go wash your hands? It's easy to think that something so small that we can't see it, can't harm us. Why, germs like *bacteria* and *viruses* are so very tiny that you'd have to have thousands of them on your hands before you could hope to see them, and you wouldn't want that. That many germs could make you very sick, indeed!

But then, not all germs cause illness. Milk cows and rattlesnakes are quite different from one another, but both are animals. Just as one animal can be so different from another, there are many different types of germs, too. Some 'good' bacteria actually live inside you and help to digest your food! Other types of helpful bacteria are used to make delicious cheeses, creamy yogurt, and cocoa for hot chocolate. Those bacteria certainly make our lives more pleasant.

On the other hand, not-so-friendly bacteria can cause painful strep throat, and other serious illnesses. And miserable colds, flu, and even warts are caused by viruses. These are germs that no one wants to have around. Nevertheless, bacteria and viruses are almost everywhere you look. You just can't see them.

Germs float through the air when a sick person coughs or sneezes. Then

Science 2 *for Little Folks*

those germs may *infect,* or 'plant' themselves in, someone else. It is possible for one sick person to make hundreds of other people sick! You can see why a considerate person covers his mouth when he coughs, covers his nose with a tissue when he sneezes, and stays home when he is not feeling well. That way he can get better without spreading his germs to others.

Germs are spread by touch, too. During cold and flu season, you will want to wash your hands even more carefully than usual. You will also want to keep your hands away from your nose and mouth. Since most germs grow best where they can find moisture, warmth, and food, they would be quite comfortable in your mouth, wouldn't they?

But how can one little germ make a person so sick? One reason is that the one little germ can quickly grow to two, then four, then eight, then sixteen and so forth. Here is a good way to understand how fast bacteria grow. Think of dropping one marble on your bedroom floor when you get up in the morning. Now, if you stepped on it with your bare foot, the hard glass ball would make your foot hurt. But it is easy to stay away from just one marble.

Imagine, however, that while you were doing your schoolwork, the marble quietly turned into two marbles, then four, then eight. If the marbles grew like bacteria do, by afternoon your bedroom floor would be completely covered with those round pieces of glass! There would no place to walk without stepping on them. Ouch! Worse, by bedtime your whole house could be filled, top to bottom, with marbles.

Now, you remember that germs are so small that you can't see them without a microscope. But once they become warm and cozy inside your body, they spread. Their numbers grow rapidly. It doesn't take long for a few bacteria to grow to many, and then you feel sick, sick, sick.

Let's think back now to the 'good' bacteria. They grow and spread, too, but in ways that make our lives better. Wouldn't it be lovely if, a little like good bacteria, we could spread the love of Our Lord and Lady to others? They might 'catch' Christian joy from us! If each of us could 'infect' one or two people every week, and then three or four more people 'caught' the Faith from them, it wouldn't be long before the whole world was filled from top to bottom with the love of Jesus. Shall we give it a try?

"From the beginning, the first disciples burned with the desire to proclaim Christ: 'We cannot but speak of what we have seen and heard.'"
—CCC 425

"Declare His glory among the nations; His marvelous works among all the peoples!"
—Ps. 96:3

Questions

1. Name two types of germs.

   _____

   _____

2. Tell three things a sick person can do to help keep from spreading his germs to others.

   _____

   _____

   _____

   _____

   _____

3. Most germs grow best where they can find what three things?

   _____

   _____

   _____

   _____

Geology

Minerals and solubility

Salt of the Earth

Do you own a strong magnifying glass or perhaps a microscope? I hope so, because they can help you to discover so many exciting things about God's handiwork. For example, try this: get a flashlight, a magnifying glass, and a shaker of salt. Turn the flashlight on and set it on its side on the table. Sprinkle a little salt on the palm of your hand. Hold your hand so the light from the flashlight shines on the salt. Now examine the salt with your magnifying glass. Can you see sparkling cubes of salt? These cubes are called *crystals*. Salt crystals form in the shape of a cube, but crystals are found in many other shapes and forms.

Most precious gems, like diamonds, emeralds, and rubies, are also mineral crystals. *Minerals* are usually found in the ground. We generally think of them as rocks, because rocks are made up of a variety of minerals. Salt is just one of many minerals.

You probably know that, while diamonds are quite expensive, a small box of salt can be purchased for less than a dollar. But can you guess which is really the more valuable of the two? It's the salt! [Do you think that people sometimes value the wrong things? That is a good question to think about, too.]

Do you know why salt is more valuable than diamonds? It's because you can live without diamonds, but you can't live without salt. Salt is found in our blood and tears; our bodies must have salt if we are to stay healthy.

One interesting characteristic of salt, which makes it easy for our bodies to use, is its **solubility**. Solubility is the ability of a substance to *dissolve*, or to blend with a liquid so completely that it disappears. When a large amount of salt is dissolved in water, the salty water is called *brine*. Brine has been used since ancient times to preserve foods such as fish and ham and pickles. The salt in brine slows the growth of harmful bacteria which makes food spoil.

Finally, salt makes food taste good! Try eating a few bites each of salted and unsalted potato. Which do you think tastes better?

In Matthew 5:13, Jesus calls us the 'salt of the earth.' How can we be salt? Well, we can add 'flavor' to life by our godly speech and actions and, with God's help, *preserve* ourselves and others from sin, which can 'spoil' the soul. But first, like a perfectly dissolved salt solution, all our thought, actions, and everything that we are must be completely 'absorbed' in Our Lord. In that way, it is no longer we who are visible, but Jesus.

Christ be with me, Christ within me,
Christ behind me, Christ before me,
Christ beside me, Christ to win me,
Christ to comfort and restore me.

Christ beneath me, Christ above me,
Christ in quiet, Christ in danger,
Christ in hearts of all that love me,
Christ in mouth of friend and stranger.
—St. Patrick's Breastplate

Questions

1. Why is salt more valuable than diamonds?

_____

_____

_____

_____

Science 2 *for Little Folks*

2. When a substance blends so completely with a liquid that the substance disappears, we say that it has:

_____

_____

3. Salt, diamonds, and rubies are all types of:

_____

_____

Experiment!

Let's test for solubility! For this experiment, you will need four clear glasses, a set of measuring spoons, a spoon for stirring, and 3 tablespoons each of uncooked rice or wheat hot cereal granules, sugar, salt, and sand or dirt. [Be sure to use a separate glass for each material that you test.]

First, fill a glass with water, leaving enough room at the top so the water won't spill when you stir it. Now add the 3 tablespoons of cereal granules and stir vigorously for about a minute. Has the cereal dissolved?

Fill another glass with water and add the 3 tablespoons salt. Can you see the salt crystals in the bottom of the glass? Stir the salt in the water for about two minutes, until all the salt has *completely* dissolved. Do you think there is still salt in the water? Taste the water and see. Now put two tablespoons of the salt solution on a saucer and set it on the windowsill until all the water evaporates. [It will probably take a few days.] What is left on the saucer after the water evaporates?

Were the dirt, sugar, or sand soluble? Are all materials soluble? Did God have a good idea when he decided to make some things *insoluble* [not able to dissolve]? What would happen if your boots and raincoat were soluble?

42 Science 2 *for Little Folks*

Chemistry

Elements, compounds; molecules, atoms

God's Building Blocks

Have you ever made and decorated a card, all by yourself, perhaps for a friend who was ill? Did you color and cut out pretty designs and then paste them to the front of the card? Isn't it fun to create new things from paper and wood and clay, especially when your creations help someone or make them happy?

Grown-ups like to create, too. Perhaps someone in your family likes to sew or build birdhouses or bookshelves. But no matter what is being made, even grown-ups have to start with materials. Not even Mom can sew a dress without fabric or thread! Man can use his God-given imagination and talents to make many different things, but all are made with materials that God created first, from *nothing*!

Did you make the paper used in your cards? No, it probably was made with wood pulp, which came from the trees that God made. Trees are made of even smaller 'building blocks' called *atoms* and *molecules*, which are so tiny that they can only be seen with the strongest of microscopes. With toy building blocks or little plastic bricks, you can make many different things like buildings and roads and even cars and airplanes. God also has used the 'building blocks' which He created, the atoms and molecules, to make everything else that He made. [Do you know what existed before He created the atom? Only God, Who always has been, and always will be!]

If you broke a piece of iron into the tiniest piece of iron that could exist, you would have one *atom* of iron. Atoms are the very smallest of 'whole' building blocks.

Next in size to the atom is the *molecule*. Atoms are put together with other atoms to make molecules, another kind of 'building block.' A molecule made up of all one kind of atom is called an *element*. Iron, for example is an element because it is just made up of atoms and molecules of iron. But if different kinds of atoms are joined together, they form a *compound*. A compound is a mixture of different kinds of atoms that makes something completely new.

One way to help us understand elements and compounds is to think about eggs and cakes. Do you like to cook? Have you ever made scrambled eggs? To make scrambled eggs, you break an egg into a pan and then mix in another egg, and you still have eggs. If you add three or four more eggs and then cook them all, you will have a breakfast of nothing but eggs. The eggs would be a little like an *element*. They would be nothing but eggs and look and taste like eggs alone.

But if you beat some eggs in a bowl and then mixed in sugar, butter, vanilla, flour, and a few other things and cooked them, you would have something new. It would not look like eggs, nor would it look like butter, or flour or any of the other individual ingredients. It would be a mixture of all of those things, and look like something totally different than all of them. It would be a cake! A *compound* is a little like the cake, made up of different kinds of atoms mixed together to make something new.

So many things can be made with God's building blocks, atoms and molecules. Your shirt may be made from cotton and polyester, two very different fabrics. But the shirt is still made of atoms and molecules that God created. Computers are made from metal and plastics, which are also made of God's atoms and molecules.

Because Our Lord gave us an intellect to learn about Him from what He has made, we can study and think and invent new things. When we are willing, He inspires us by the Holy Spirit to make things to benefit ourselves and others. Finally, He provides the material necessary for the task. In a sense, man has never made anything completely new, because he is just using materials that God made first. Everything that man has ever invented has the same basic 'recipe': God's gift of intellect, God's inspiration, and God-created atoms, molecules, elements, and compounds.

"To create is to bring a thing into existence without any previous material at all to work on. 'In the beginning God created the heaven and the earth.'"
—St. Thomas Aquinas [13th century]

"God...made all things by His Word, and fashioned and formed that which has existence out of that which had none."
—St. Irenaeus [2nd century]

Questions

1. Name the very smallest of building blocks.

   _____

   _____

2. A molecule made up of one kind of atom is called an:

   _____

   _____

3. Different kinds of atoms are joined together to form a:

   _____

   _____

4. All that has ever been, or ever will be made, began with material created by:

   _____

   _____

Physical Science
Freezing and solids

Freezing and Solids

Would you be surprised if, at breakfast one morning, you couldn't drink the milk in your glass because it had suddenly become solid? Or if you discovered that you could never again enjoy an icy cold glass of juice on a hot summer's day, because all liquids had become solid and had to be heated before they would turn into liquid again?

We usually think of water as a liquid, but when the temperature of water is lowered to 32º F [0º C], it becomes a solid—ice—by a process called *freezing*. When we think of 'freezing,' we think of cold and ice, but freezing doesn't always mean 'cold.' 'Freezing' really means *the point at which a liquid becomes solid*. For example, your color crayons 'freeze' at about 90º F [32º C]. That means they are solid until it is about 90º F [32º C]. Above that temperature, the crayon melts and becomes a liquid. [Have you ever left a crayon on the sidewalk on a hot summer day? What a mess!] Do you think it would be difficult to color a picture if your crayons, like water, stayed liquid until the temperature got down to 32º F [0º C]? You would have to wait for a good frost, bundle up in your warmest clothes and mittens, and go outside to do your coloring! If you brought your colors back inside, they would again melt into a puddle, unless you kept them in the freezer.

Like crayons, the gasoline for your car also becomes solid, or freezes, at a different temperature from water. The point at which gasoline becomes solid is so low that it is almost impossible to make gasoline freeze! That is a very good thing, because can you think what might happen if gasoline became solid at the same temperature as water? That's right! You could never drive your car in the winter, because the gas would freeze solid in the tank. It would be hard to travel to visit with grandparents at Christmas time, and people who lived in cold areas would all have to walk to Midnight Mass. All the trucks that deliver food to the stores, and ambulances and fire trucks, would be stuck until the weather warmed up.

Even things which seem always to be solid, like iron stoves, have a freezing point! Iron remains solid up to about 2,795° F [1,535° C]. That is, at that extremely HOT temperature, iron melts; below that temperature, iron stays solid. What if God had decided to make iron melt at the same temperature as crayons? Could we make our stoves from metal? What would happen when your mom decided to use the oven to bake a birthday cake for you? Why, the stove would melt!

Can you see why our very wise God would create so many different materials with such different freezing points?

"Bless the Lord, ice and cold, sing praise to Him and highly exalt Him forever."
—Dn. 3:49

"The very order, disposition, beauty, change, and motion of the world and of all visible things proclaim that it could only have been made by God, the ineffably and invisibly great and the ineffably and invisibly beautiful."
—St. Augustine, The City of God, 11, 4, 2

Question

1. The point at which a liquid becomes solid is called what?

_____

_____

Experiment!

Water, juice, and soda pop are all *liquids*. If you want to make them into *solids*, you can put them in the freezer! Pour about two tablespoons of oil, juice, water, milk, and vanilla flavoring into small, disposable plastic cups. [Use a different cup for each liquid.] Put the cups into the freezer for about four hours. Then check, by poking the contents with a fork, to see what surprises you find! [And, if you freeze a few extra cups of juice, you will have some delicious popsicles!]

Science 2 *for Little Folks* 47

Astronomy

Solar system:
Sun and moon

Light from Light

Have you ever been caught in the dark when the power suddenly went out? Did you stumble around and bump into furniture while you were looking for a flashlight? Did your parents light a few candles, and maybe let you carry one for a moment? [You have to be very, very careful with candles, and only use them with your parents' permission!] Without a light, it can be a little bit scary when you can't tell what is ahead of you in the dark, can't it? Thanks be to God, He is always with us and sees clearly, no matter how dark the night.

What a difference to have the power go out on a sunny day! The light of the sun is so bright, we hardly notice that the electricity is off. Did you know that, if you could hold the sun in your hand to light your way, it would be as bright as about 2.5 billion billion billion candles? Of course, you cannot hold the sun, because it is a huge fiery ball of gas. To help you 'see' the size of this great 'fire in the heavens,' ask your parents if you may use a tape measure. Measure out a line 109 inches long. That would represent the diameter, or a line straight through the middle, of the sun. Now measure just one inch. That inch would represent the diameter of our planet earth, compared to the sun. Even though the sun is 93 million miles away from the earth, its great size and flaming gases warm us and light our way. Without the sun, life could not survive on our planet.

In contrast, if the diameter of the sun were 109 inches, the diameter of the moon would be only 1/4 inch! Yet, a full moon floods the night with its gentle glow. However, unlike the sun which makes its own light, the moon has no brightness of its own.

Do you wonder where the moon gets its light? From the sun! Even after the sun sets, its brilliant rays continue to shine brightly into space. Our Lord created the little moon, with no light of its own, to receive the light from the sun and bounce, or *reflect*, it down to earth. The next time you enjoy the splendor of a moonlit night, remember that you are really seeing the light of the sun, reflected from moon to earth.

Sometimes we cannot see the moon, but it is still there *orbiting*, or circling, the earth. Neither can we see the sun at night, because the earth is constantly turning. At night, our side of the earth turns away from the sun. During the day, our side of the earth faces its brilliance. But even though there are times that we can see neither sun nor moon, they are always there, faithfully performing the tasks that God assigned them.

God is the Maker of all that we discover in science. He delights in showing Himself to us through His creation. In this case, it is the sun and the moon which help us better to understand Our Lord and Lady. Our Lord is the source of all grace, goodness, and light; we cannot have life apart from Him. His creation, the sun, is the source of all light. It is absolutely necessary for life on earth.

As the moon orbits close by and reflects the light of the sun to earth, so Our Lady stays close to us, perfectly reflecting God's will. She showers us with light and grace which come from Our Lord. When we are struggling with the darkness of sin, the glorious 'Son-light' of Jesus shows us the true way.

"Thou art, O God, the life and light
Of all this wondrous world we see,
Its glow by day, its smile by night
Are but reflections caught from Thee—
Where 'er we turn, Thy glories shine,
And all things fair and bright are Thine."
—Anonymous

Questions

1. When light bounces from one object to another, we say that it:

   _____

   _____

2. How far is the sun from the earth?

_____

_____

3. Where does the moon gets its light?

_____

_____

Experiment!

For this experiment, you will need a flashlight and a hand mirror. Turn the flashlight on, and set it on a table or counter so that the beam is directed at the ceiling. Turn off all lights. Hold the hand mirror above the flashlight so that the beam reflects off the mirror.

- Can you make the light reflect onto the floor?
- Is the pool of light on the floor coming from the mirror or the flashlight?
- Using the mirror, can you direct the light onto the walls and other parts of the room without moving the flashlight?
- Where does the light come from?
- Does reflected light feel warm? Why or why not?
- Can you make a list of other light sources, such as light bulbs?
- How many can you think of?
- Do the light sources you have listed also produce heat or warmth?
- Can you think of other ways to reflect light? [Hint: have you ever seen bright spots on the kitchen ceiling when the sun shone through the window onto the surface of a pan of water?]

Astronomy

Solar system: Meteors

Falling Stars

Have you ever seen a 'falling star' streak across the night sky? [Some people call them 'shooting stars.'] Did you know that they really aren't stars at all? Their real name is **meteors**, and that is what this science story is all about.

Meteors, smaller than a grain of rice, are really tiny pieces of rock and dust. You could put several of them on the tip of one of your fingers, if you could catch them before they fall.

Giant stars keep their places in the heavens, millions of miles away, but tiny meteors fall through the sky as close as 40 or 50 miles above the earth. But you can't catch one, because it burns up before it reaches the earth!

As the meteors fall, **friction** with the air makes these little specks so hot that they begin to burn. [Do you know what friction is? Rub the palms of your hands together, back and forth as fast as you can. Do this about 60 times. Do your hands feel hot? The rubbing is a type of friction, which can make things so hot that they burn.] When the tiny meteors burn, they glow so brightly that you can see them easily against the dark sky!

Every year, God causes the earth to pass through several 'meteor showers,' during which you can watch these glowing streaks of light. One of these showers, the 'Geminids,' can be seen from about December 13-15th, in the southern sky. However, my favorite 'showers,' the 'Perseids,' can be viewed from around August 11-13th. By looking into the northeast sky, you can see as many as 80 or 90 meteors each hour. [They are best seen away from city lights, where the sky is darker.] I like to lie down on the warm grass in the back yard about two hours after sunset and stare into the heavens. Each bright 'falling star' that I see blaze across the sky reminds me that Our Lord can take something as tiny and simple as a piece of dust and turn it into something spectacular for His glory. Would you like to offer yourself as a 'speck of dust' in God's hands?

"Remember, man, that you are dust, and to dust you shall return."

"The heavens declare the glory of God..."
—Ps.19:1

Questions

1. What are two other names for 'falling stars'?

   _____

   _____

2. When can you watch the Perseid showers?

   _____

   _____

3. What are 'falling stars' made of?

   _____

   _____

4. Do you know the meaning of *spectacular*? Can you find it in a dictionary?

   _____

   _____

   _____

   _____

Astronomy

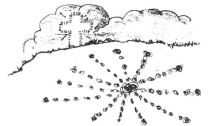

Searching the Heavens

Who is God? Where is God? For countless ages people have asked, and still ask, these questions. Often, their search for the answers begins as they stand staring in amazement at the night sky. Who set the twinkling stars in their places? And why does the sun disappear each day in the west, and then faithfully reappear the next morning in the east? Truly, the heavens tell the glory of God.

Long ago, people in different parts of the world tried to understand God and His heavens by building simple **observatories**, or places to study the stars and planets. One such place, called Stonehenge, was built in England about 4,000 years ago. We don't know exactly how or why Stonehenge was made. But we do know that the gigantic stones that make up this 'observatory' are set up in a special way. At the summer **solstice**, the longest day of the year, one can watch the sun rise exactly between two of the stones. Some *astronomers*, or people who study the stars, moon, sun, and planets, think that Stonehenge was used to keep track of the movements of the sun and moon. The people who built Stonehenge did not know the God Who made the universe, but they probably were trying to find Him by studying His heavens.

In North America, there are stone 'observatories' of a different type. Located in Canada are the Majorville

Science 2 *for Little Folks*

Cairn [in Alberta], and the Moose Mountain Cairn [in Saskatchewan]. Another, the Bighorn Medicine Wheel, can be found in the state of Wyoming. Like Stonehenge, these 'observatories' are also made of rocks. Native Peoples built these 'observatories' perhaps two or three thousand years ago. They set stones in patterns that look like big wheels, some about eighty feet wide. Like certain stones at Stonehenge, the rocks at Bighorn Medicine Wheel point to the place where the sun rises at the summer solstice. Other stones seem to point to the constellations of Orion and Taurus. Those who set the rocks in place were surely trying to understand our Creator and His wonderfully designed universe.

A modern observatory, located in Mount Graham, Arizona, was started by the Vatican! [You probably knew that the Vatican is the 'home' of Jesus' One, Holy, Catholic, and Apostolic Church.] This observatory is different from Stonehenge, the Cairns, and Medicine Wheel, because its makers know their Maker. They already understand that our Almighty God and Creator is the One Who made and then set the sun and moon and stars in their places. Astronomers at the Vatican Observatory are still learning new things about the sun and moon and stars. But they understand that God, the Creator of the universe, already knows its secrets.

Who is God? Where is God? The wonderful answers to these questions we learn from Church teaching, and also from what we can observe. We can see that the whole universe—the sun, the moon, the stars, the planets—was made by Someone even greater than the universe itself. That Someone is God, and He is everywhere. He is even bigger than the universe that He created, but He can live inside your heart. What a blessed mystery!

"The heavens are telling the glory of God, and the firmament proclaims His handiwork."

—Ps. 19:2

"No one has ever seen God. Yet if we love one another, God dwells in us..."

—1 Jn. 4:12

Questions

1. Another name for places built to study the stars and planets is:

   _____

   _____

2. Another name for people who study the stars, moon, sun, and planets is:

   _____

   _____

3. The longest day of the year is called the summer what?

   _____

   _____

4. Where is the modern Vatican Observatory found?

   _____

   _____

5. Find the constellations Orion and Taurus on a star map. In the space below, draw these constellations.

Astronomy

Solar system:
Earth's rotation

From the Rising of the Sun...

Have you ever watched on television as the Holy Father celebrated Midnight Mass at Christmas? Perhaps you wondered how it could already be Christmas in Rome, since it was still Christmas Eve at your house. If you have a flashlight and a world globe, you can find out the answer. [If you don't have a globe, maybe you can make one from a beachball or basketball.]

Take your globe and flashlight into a dark room. Set your globe on a table or other flat surface. Now hold your flashlight so it shines on the globe. Your globe, or model of the earth, will be light on the side facing the flashlight. It will be dark on the side that faces away.

Now, holding your flashlight, or 'sun,' about two inches away from the globe, turn your 'earth' slowly to the right, so the 'sunlight' hits the east coast of North America. Stop for a moment. Can you see that, while the east coast is lit up from Newfoundland to Florida, it is still dark on the west coast? Turn your globe slowly again. As your 'earth' turns, you can watch the 'sun' come up on the west coast, too. As you keep turning the globe slowly, you can see that it becomes dark first on the east coast, and then on the west coast. This is just the way that our earth turns, with one side facing the sun and one side facing away from the sun. If you are up early enough, you may be treated to a beautiful sunrise if you look

56 Science 2 *for Little Folks*

to the east. In the evening, if you face west you may see a brilliantly colored sunset. But really, it is not the sun that rises or sets to make our days and nights. In space, the sun shines all the time! Instead, it is the way that God has made the earth spin around and around that divides our hours into night and day.

On a piece of paper about the size of a dime, draw a little stick person. We will pretend that the little paper person is you! Tape the person on the globe where you live on the 'earth.' Turn the globe so the light is shining on Rome, Italy. [Remember that Rome is where our Holy Father celebrates Christmas Mass.] When it is light in Rome, is it light where you live, too? [Be sure to hold the flashlight about two inches away from the globe.] Do you see that the 'sun' comes up first in Rome, and later where you live? If the sun comes up first in Rome, it also means that the sun will go down first there, too. That is, it will be night in Rome before it is night at your house.

You can see from looking at your globe that, when it is dark on one side of the world, it light on the other side. So, not long after North American children have snuggled under their covers for a good night's sleep, children in Asia are probably getting ready for lunch. Wouldn't it be funny if 'noon' meant lunchtime for children in Asia, but bedtime for you?

It would be confusing if, all around the world, our clocks said the same thing at the same time. Some might be lunching on a steaming bowl of chicken noodle soup at noon while others were crawling into bed for the night, at noon! About one hundred twenty years ago, others thought that would be odd, too. So they created what are called *time zones*. Each time zone has a name. In North America, beginning in the east at Nova Scotia and ending in the west with Alaska, the zones are *Atlantic, Eastern, Central, Mountain, Pacific,* and *Alaska*. Moving from west to east, each zone adds one hour of time to the clock. Time zones like these make it so that noon can be lunchtime, and midnight can be sleeping time, no matter where you live in the world.

For example, it may be light at your house, so it is still daytime. But the sun has probably gone down in Rome, so it is night there. The people in Rome are in a different time zone from yours. They live in a time zone 'ahead' of the time zone in which you live. [Can you see this from your globe?] But they still get up when it is morning where they live, and they still go to bed when it is nighttime where they live. That is why the Holy Father can celebrate Midnight Mass before it is midnight where you live. He can celebrate Midnight Mass in Rome, because it is already midnight there.

Turn your globe slowly around once again. Since Our Lord has made the world to keep spinning all the time, do you see that, somewhere in the world, the sun is always coming up? Can you see that at the same time, somewhere in the world, the sun is always going down? I will tell you something wonderful about this.

No matter where you live in the world, there is no better way to start the day than with prayer. And the very best prayer that we can pray is the Holy Sacrifice of the Mass. Holy Mass makes Christ present to us. It also joins us with all of our other brothers and sisters in heaven and everywhere on earth, in the one Body of Christ.

Think of this: Somewhere in the world, the sun is always rising; it is always daylight somewhere in the world. So too, the Light of Christ is always with us, especially in Holy Mass. Holy Mass never ends, for just as it finishes in one place, it starts in another! Every hour of the day and night, this great Sacrifice of Jesus' love poured out for us is being offered somewhere around the earth. What glory, to have the living presence of Jesus Christ, always with His world. Truly, He never leaves nor forsakes us. Alleluia!

"I will not leave thee, neither will I forsake thee."

—Heb. 5:13 [Douay-Rheims]

"For from the rising of the sun, even to its setting, My name is great among the nations; and everywhere they bring sacrifice to My name, and a pure offering."

—Mal. 1:11

Questions

1. If it is daytime where you live, is it also daytime on the opposite side of the world?

   _____

   _____

2. If you want to see the sun rise, do you look to the east or to the west?

   _____

   _____

3. Look in a telephone book and find your time zone. Write its name below.

   _____

   _____

Earth Science

Weather: Precipitation

Merciful Rain

There is nothing as refreshing on a blistering August day than cold, clear water from the kitchen faucet, don't you think? As you run that bubbling, cool liquid into your glass, do you ever stop to think where the water comes from? Perhaps your water is piped to you from a well, or maybe from your city's water system. But how does the water get into the well or the water system in the first place?

Precipitation is what we call the way that God provides water for His earth. Some kinds of precipitation are rain, hail, and snow. All of these types of precipitation start as water and end up as water. But water isn't always a liquid. Water can become a gas that you can't see, called *water vapor*. Water can also become a solid, called ice. The water that God sends for you to drink may change from a vapor to a solid to a liquid before it travels through your faucet!

Rain, hail, and snow begin in clouds made up of tiny water droplets or pieces of ice called *crystals*. Raindrops start with a tiny piece of dust or ice at their center. That little center piece collects tiny droplets inside the cloud. The new raindrop gets bigger and bigger until it falls to earth, perhaps to help fill your favorite cup.

Snow starts with a tiny piece of dust or ice at the center, too. The itty bitty piece at the middle of the snowflake gathers water vapor that turns right into ice crystals. As more ice crystals stick together, the snowflakes get bigger and bigger until they drift to the ground, blanketing city and country in soft, white stillness.

Hail is icy, like snow, but God has a completely different way of making it. Hail always is made in huge thunderclouds, and it begins as rain. Inside the thundercloud, the raindrop is tossed about. The raindrop passes through freezing air and turns into ice inside the cloud. Then the frozen drop begins to fall. As it falls down, down, down through the giant cloud, it picks up water droplets. But then the moving air in the thundercloud tosses the wet ice-ball back up into the freezing part of the cloud. That makes the ice-ball grow into an even larger ball of hail. After a time, the hailstone is so heavy with layer after layer of ice that it rockets down onto a field or perhaps bounces loudly off the roof of your car.

Whether from rain, or snow, or hail, precipitation brings water to fields, rivers, and lakes. The water from those fields, rivers, and lakes eventually finds its way to your kitchen faucet. Now, you may live in a place that seldom has snow. But you could still be enjoying water that came from snow!

Have you ever filled a drinking glass with nothing but ice cubes from your freezer and then, as the ice slowly melted, sipped the cool water from the melted ice to quench your thirst? Sometimes the ice takes two or three hours to melt. Because the ice melts slowly and makes a little bit of water at a time, you can enjoy those sips of icy goodness for a long time.

God has a similar system for supplying water to places that don't get much rain in the summer. Instead of a freezer, God uses tall, snow-covered mountains to store up icy snow. In some mountain ranges, like the Rocky Mountains that stretch north and south across Canada and the United States, peaks often receive thirty or forty feet of snow each year. That is nearly enough snow to bury a Ferris wheel! Even after winter has passed, the cool mountain air allows the snow to melt very slowly. Melting snow slips into creeks and rivers that carry needed water to farms and cities far away. Some of those farms and cities don't get much rain in the summer. Some of them don't even have snow in the winter. But the people who live there are happy to drink and swim in the water that came from God's far-away frozen mountains.

Then there are places that have no mountains to store their snow for summer drinking and swimming. They are too far from rivers to use river water, either. But God has planned a way for them to have water in the summer, too. In flat lands like Iowa and Manitoba, heavy summer rainstorms and hailstorms supply water for drinking and farming.

Whether falling as snow, hail, or rain, precipitation brings life-giving water to all of God's children. Scripture tells us that God causes the rain to fall on the just and the unjust. That means that He generously gives rain to those who love Him, and even to those who do not love Him. In the same way, God showers His mercy on all people.

God loves and cares for all His children, and He asks us to do the same. We can imitate Jesus' love by being kind to and praying for those who don't love us. Let us praise and thank Him for the water He sends from the Heavens, and for His unending care.

"Love your enemies and pray for those who persecute you, so that you may be sons of your Father...for he ...sends rain on the just and on the unjust."
—Matt. 5:44-45

"As for me, I would seek God, and to God would I commit my cause; who does great things and unsearchable, marvelous things without number: he gives rain upon the earth and sends waters upon the fields......"
—Jb. 5:8-10

Questions

1. What do we call the way that God provides water for His earth?

   _____

   _____

2. What do we call water that has become a gas that you can't see?

   _____

   _____

3. What do we call the tiny pieces of ice that are found in snow clouds?

   _____

   _____

Earth Science

Weather: Lightning

Lightning!

When you hear thunder's deep rumble and see lightning's blinding flash, are you afraid?

Are you afraid of a hot stove? Good for you! Sometimes it is good to be a little afraid of things that might hurt us, so we will be more careful. So it is not a bad thing to be a little afraid of thunder and lightning. But did you know that lightning is helpful, too?

Lightning is a type of electricity, similar to what we use to light our homes, and even acting in some ways like the battery you use in a flashlight. If you look at a battery, you will see a little '-' on one end and a '+' on the other. These are called *negative* and *positive charges*. Lightning happens when a **HUGE** number of negative charges and positive charges meet in the air outside.

When the positive and negative charges rush together to make lightning, the flash is so hot that it makes a big sound wave in the air. You hear it as booming thunder. But the lightning flash does something else that is another of God's miracles.

Plants need food, just as you do. One of the types of food that they need is called **nitrogen**. There is a lot of nitrogen in the air, but guess what? The plants can't use what is in the air. [It is a little like having your Mom hand you a raw egg for breakfast—yuck! But if the egg was changed into another form by cooking, yum!] When lightning passes through the thundercloud, it changes the nitrogen into a form that the plants can use. When it rains, this 'food' falls to the ground, where the plants use it to grow strong and healthy. Can you see why some of the states that get the most lightning storms grow some of the very best food crops in the United States?

Everything in Creation Our Lord planned very carefully, even lightning. When we see the flash of lightning and hear the **BOOM** of thunder, we remember that they come from our All-Powerful God; He has created them for our good.

"The Lord is my helper, I will not be afraid..."

—Heb. 13:6

*"Full of glory, full of wonders,
Majesty divine!
Mid Thine everlasting thunders
How Thy lightnings shine!"*

—Fr. Frederick Faber

Questions

1. Lightning happens when what and what meet in the air?

   _____

   _____

2. What is the name of the 'food' that helps plants to grow?

   _____

   _____

Science 2 *for Little Folks*

Earth Science

Geology: Weathering

Weathering the Storm

When my brother and I were children, it seemed that the happiest way to spend a summer's day was at the beach. There, low waves broke quietly over our wriggling toes and spread, sparkling, onto sun-warmed sand. Gentle breezes softly stroked the tuft grasses lining the beach's edge. How pleasant and peaceful it was to play next to the sea!

But a visit to that same shoreline in mid-winter showed a far different picture! Now, gigantic, stormy waves smashed high against sandstone cliffs. Fierce winds, carrying sand ripped from the beach, buried torn grass stems under mounded graves. From the safety of the car, my brother and I watched the roaring ocean. The beach was no longer a pleasant place to play.

When we think of storms, we often think of danger. Rainstorms can cause flooding. Windstorms can blow down trees and damage houses. Freezing ice storms break branches and make driving, and even walking, risky.

Yet, Our Lord makes good use of wind and rain, flooding and freezing. For example, farmers cannot grow food crops in hard rock. But after centuries and centuries of wind and rain, flooding and freezing, stony soil may turn into plant-friendly fields. We call these changes to rock and soil, caused by the weather, *weathering*.

Let's pretend that we've lived for thousands of years next to a rock, or boulder, that started out as big as you are. There was no soil on or around the rock, so no grasses or bushes or trees could grow there. But then came an ice storm, and the weathering of the boulder began.

A tiny crack appeared in the big rock. Some of the ice melted, and a little water dripped into the crack. At night, the water froze again. You may know that, when water freezes, it *expands,* or grows larger. As the water expanded in the crack, the crack grew bigger.

Time passed, and a windstorm beat against our rock. The beating winds scrubbed away tiny, tiny pieces of rock. Some of these tiny pieces rolled down into the crack. Then came summer rains, washing more itty-bitty grains of rock into the crack. Summer rains also brought muddy flood waters, which dropped sand and soil on the rocky ground beneath the boulder.

Fall arrived, and now the little split in the boulder was just big enough that a few blowing leaves landed inside. There, they rotted, or composted, and mixed with little pieces of rock in the crack. Soon the weather grew cold again and froze the water and the grains of rock in the crack. The crack spread wider. Now there was room in the crack for even more tiny bits of rock, and more leaves. Soil was forming in the split in the boulder! And one windy day, seeds blew into the crack and onto the thin layer of soil beneath the boulder.

With the coming of spring, the seeds began to sprout. In the boulder's crack, the tender sprouts grew into small plants. Their roots filled the crack in the rock and, ever so slowly, pushed against the hard stone sides. Now the split became even larger! Over time, further weathering broke the rock into several chunks. Some of the chunks rolled to the dirt below. When the chunks hit the ground, some smashed into yet smaller pieces. After thousands of years of weathering, the mighty rock was not so mighty. Instead of a gigantic boulder surrounded by bare rock, now there stood a small mound of broken stone, circled by shady trees and leafy bushes and thick grass.

You can see that the weathering of wind, rain, floods, and freezing can sometimes be a good thing, but it can also cause damage. Sometimes our souls face spiritual weathering, too. When temptations first begin, they often appear as harmless as the gentle breeze on a sunny beach. But before long, those temptations may start to storm against our souls. If we don't get out of the storm, we may find that temptations turn into venial sins, which wear away the grace in our souls. Of course, venial sins can lead us into the serious danger of mortal sin.

Now, a rock can't move away from the weathering, because it isn't a living, moving thing. It has no mind or free will, as people do. It cannot choose to please God and grow in holiness. But people can choose to turn away from dangerous storms of temptation and take shelter in the sacraments!

We can also decide to stay away from, or avoid, the near occasions of sin. For example, if Mom says not to touch the plate of chocolate chip cookies on the counter, we don't hang around and sniff the cookies. That would be putting the temptation right under our noses and might make us want to disobey and snatch a cookie. No, we get out of the kitchen and find some good activity to keep us busy and happy and far from those baked temptations.

If you are battered by temptation, remember that God is always beside you. Cry out to the Lord. When storms threaten to wear away His grace in your soul, He will help you stand strong.

"...they cried to the LORD in their trouble, and He delivered them from their distress; He made the storm be still, and the waves of the sea were hushed."
　　　　　　　　　—Ps. 107: 28 & 29

Questions

1. Another name for changes to rock and soil, caused by the weather, is:

   _____

   _____

2. When water freezes, does it take up more space or less space?

   _____

   _____

3. Do rocks have free will?

   _____

   _____

4. Find out about free will. Write the definition of free will below.

_____

_____

_____

5. Find an empty plastic film container [35 mm]. Fill a bowl with cold water, and hold the container and its lid under water. Let all the air out of the lid and the container. While the container is still under water, push the lid on tightly. Set the film container in a baking pan. Put the pan into the freezer overnight. Check the container the next morning. Tell about what happened and why.

_____

_____

_____

_____

_____

_____

_____

Science 2 *for Little folks*

Earth Science

Geology: Volcanic, Metamorphic rock

Rock My Soul

"I believe in God, the Father Almighty, Creator of heaven and earth…"

Are those words familiar to you? I'll bet you say them every day, when you pray the rosary. Of course, those words begin the Apostles' Creed, the first prayer of the rosary after you make the Sign of the Cross.

Often, when we think of God as Creator, the first 'picture' that pops into our heads is one of God creating the earth 'in the beginning.' At first, there was nothing at all. Then, God made great oceans and mountains, animals of every size and shape, trees and blooming plants, and our first parents, Adam and Eve. All those He created long, long ago.

But our mighty Father was not done with His creation. Did you ever stop to think that He is at it still? Every new plant whose tender leaves stretch toward the sky, every new soul born into the world, every new rock that forms, shows that God is still 'working' on His beautiful world.

What was that about new rocks? Did you think that all rocks were old? Let me tell you about some 'new' rocks that Our Lord is making, even as you read this story.

Deep within the earth, under the volcanoes of Hawaii, hot, melted rock pushes up, up, up. This melted rock bursts through the surface of the earth and into the air as *lava*, in a fiery volcanic explosion, or *eruption*. Sometimes the glowing lava runs down the side of the volcano like a bubbling river of hot rock, flowing to the sea. New land forms as the lava cools at the ocean's edge. Long ago, God made the Hawaiian Islands from volcanic rock. But even today, God is still using His volcanoes to make the Hawaiian Islands grow.

And Hawaii is not the only place where you can find volcanoes and new rocks and new land. Not so very many years ago, a farmer was at work in Mexico. He saw something strange happening in his cornfield. Pieces of hot rock and ashes were coming from a crack in the ground. In less than a week, where his cornstalks had once stood, there was a mountain as tall as a forty-five story skyscraper. Over the next nine years, the mountain [now called *Paricutin*] had grown to 1,345 feet tall. In that time, it had thrown rock and ash more than ten miles away.

When volcanoes first spit out rock and ash, the soil is too 'new' for plants to grow well. But, after a time, volcanic ash turns into rich earth. All around the world, from the Philippines, to Italy, to Japan, farmers plant rice, grapes, apples, and many other types of fruits and vegetables close to volcanoes. They thank God for His excellent volcanic 'dirt' that causes their plants and trees to grow so much good food.

Now, volcanic rock isn't the only kind of 'new' rock that God still makes.

Metamorphic rock is made by changing one kind of rock into another. Often, after being changed, the metamorphic rock is far stronger and even more beautiful than it was in its first form.

For example, limestone is a rather plain-looking type of soft rock. Some types of limestone, like chalk, are so soft that they can easily be broken by hand. If you wanted to carve a fine statue, you would not choose limestone to carve. If you did, wind and rain would slowly wear away your carving. No, you would want a stronger rock.

But Our Lord can turn that soft, plain limestone into a different, strong and lovely metamorphic rock called *marble*. God uses pressure, or the weight of soil and rock pushing down on the limestone, and heat from the earth, to make the change. Of course, the process takes many long years but, in the end, a more beautiful and long-lasting stone is formed. If you have ever seen any *sculptures*, or carvings, made by a man named Michelangelo, you have seen the beauty of God's marble touched by God's inspiration.

Now, rock isn't the only part of God's creation that He's still working on. Our Lord is still working on His people, too. Some of us are like 'new' volcanic soil, still in need of Jesus' hand, and a little time, to make us 'rich' so 'holy fruit' can spring from us. Others are soft and weak, like limestone, and could use a little pressure and heat to make us strong of soul and lovely of heart. May we all be like marble in His hands, ready and willing to cooperate with Our Lord as He shapes us into people who will reflect His glory.

"I will give you a new heart and place a new spirit within you, taking from your bodies your stony hearts and giving you natural hearts. I will put My spirit within you..."

—Ez. 36:26

"Creation has its own goodness and proper perfection, but it did not spring forth complete from the hands of the Creator. The universe was created 'in a state of journeying' [in statu viae] toward an ultimate perfection yet to be attained..."

—CCC 302

Questions

1. Long ago, God made the Hawaiian Islands from what?

   _____

   _____

2. Name a strong and lovely metamorphic rock.

   _____

   _____

3. Find out about the sculptor Michelangelo. Tell a little about his sculpture, the Pieta.

   _____

   _____

   _____

   _____

Earth Science

Geology: Sedimentary rock, fossils

Layers of Rock, Layers of Faith

If you've ever rolled a candy jawbreaker around in your mouth, you know what it's like to wonder what flavor your taste buds will discover next. Perhaps you removed the sticky jawbreaker from your mouth as you sucked on it, just to see whether cherry red had turned to blueberry blue or banana yellow. Did you ever smash one of the candies open with a hammer so that you could see the layers and layers of different flavors and colors hidden inside? [Of course, you wouldn't want to crack the jawbreaker open by biting down on the hard, round candy. That might crack your teeth, instead!]

You may already have studied igneous rock, formed by fiery volcanoes. Maybe you also know a bit about metamorphic rock, created when one type of stone is changed into another. A third kind of stone, **sedimentary** rock, might remind you of the jawbreaker's sugary layers.

Sedimentary rock can be created when passing flood waters drop mud or sand onto older layers, or deposits, of mud and sand. Year after year, these deposits slowly pile up, one on top of the other. Sometimes the layers are different colors, a little like the changing hues of your jawbreaker. If you have ever seen the 'painted' rock walls of the Grand Canyon, you have viewed a magnificent example of our Creator's sedimentary designs.

Another sedimentary design also begins in the soft mud left by flood waters. Now and then, buzzing insects or feathery ferns or shellfish get stuck in the oozing, thick mud. As time goes by, the weight of the new mud on top pushes down on the old mud below. Gradually, the older buried mud becomes hard. The insects, leaves, and shells that were caught in the mud become part of the stone, or leave their shapes in the rock. These shapes of plants and animals are called *fossils*.

Would you like to find a fossil? Fossil imprints can be unearthed in riverbanks, on cliff faces by the seashore, or natural rock walls that have been uncovered by road building. Many walk right past fossils, not even knowing that they are there, hiding in the rock. Search for them; if you find a large chunk of sedimentary rock in one of these places, stop and take a closer look! You may have a happy surprise. Perhaps you will find a fossil seashell right on top, if you take the time to look closely and seek with all your heart. If your piece of rock is large enough, you may be able to break away layer after rocky layer, discovering new animal or plant fossils hidden beneath each one.

Often, as we become older and grow to know, love, and serve Jesus better and better, we are happily surprised to learn that there are also layers to our Holy Faith. For example, when you pray the Joyful Mysteries of the Most Holy Rosary, you *meditate,* or think in a prayerful way, of the Finding of Our Lord in the Temple. Another layer of this mystery is that you can find Jesus in the Temple, too! Do you seek the Lord with all your heart at Holy Mass, or find Him while paying a visit to Jesus, truly present in the tabernacle?

What about that Sorrowful Mystery, when cruel soldiers pressed a crown of sharp thorns upon Our Lord's holy head? Don't 'pass by' this mystery, thinking that it tells only about something soldiers did long ago. There are other surprises hidden inside; search for them! Seek with all your heart. You know that Jesus is our King. He should therefore be King of our actions and words, shouldn't He? My Jesus, may I never press upon Your head a thorny crown of meanness and bitterness, but instead present to You a golden crown of kind deeds and speech!

Perhaps you know all the Luminous Mysteries, including Jesus' first miracle. After you meditate on Our Lady pointing others to her Son at Cana, why not dig a little deeper, to the next layer? You will surely be delighted to find that she also speaks to you: 'Do whatever He tells you.'

Take a closer look the next time you pray the rosary. Ask the Holy Spirit and our Blessed Mother for the gift of understanding. They will help you uncover and understand your part in the beautiful, breathtaking layers found in the mysteries of the Most Holy Rosary. You will surely find glorious, ancient truths, presented in a new way.

"You will seek Me and find Me; when you seek Me with all your heart, I will be found by you, says the LORD..."
—Jer. 29:13&14

"Faith merely assents to what is proposed, but the gift of understanding brings some insight into the truth..."
—St. Thomas Aquinas, Summa Theologica, 2-2, 8, 5

Questions

1. When passing flood waters drop mud or sand onto older layers of mud and sand, after many years, what kind of rock is created?

_____

_____

2. Shapes of insects, leaves, or shells that are found in rock are called what?

_____

_____

3. Find out about the Grand Canyon. Tell about two things that you discovered.

_____

_____

_____

_____

_____

4. Another word for 'thinking in a prayerful way' is:

_____

_____

Botany

Seeds: Structure and germination

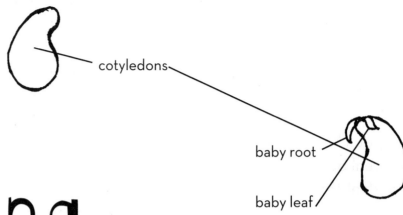

Feeding Baby Plants

Do you have a baby at your house? I'll bet that, in your home or in your parish, you often see babies being lovingly fed and cared for by their mothers. Do you think a baby could live if it didn't have someone to care for it? No way! Thanks be to God that dads and moms are part of His perfect plan. But did you know that there is a type of 'baby' that can 'take care' of itself? It is called a *plant seed!*

Our Lord gave plants a way to make their own food by using sunlight, water, carbon dioxide and a substance called **chlorophyll**. Chlorophyll is found in leaves; it is also what makes them green. So, using their leaves, plants can take care of most of their food needs. With their leaves, they can feed themselves even when they are very young. That was pretty smart of God, because seeds don't have a mom to care for them.

However, a baby plant starts out as a seed, and seeds don't have leaves! So how can the baby plant feed itself when it is starting to grow, hidden in the warm earth? How can it feed itself when it first pops out of the ground if it doesn't yet have leaves? Why doesn't it just starve to death? Let's find out!

When a bean is planted, it first sends out a root to hold the plant in place and to get water for the plant. Next, the stem for the leaf begins to grow up toward the light. Last of all, the tiny leaf appears.

With all that work of growing, the baby plant needed lots of food for energy, but where did it get it? It did not yet have leaves to make food for itself! If you have some dried beans in your house, you can look at them and find out.

First, put a handful of dried lima beans into a cereal bowl. Fill the bowl with water and let the beans sit overnight so they will open more easily. The next day, carefully slip the skin off the seed, open it, and you will see that the seed has two neat halves. These are called **cotyledons**. You will also see a tiny leaf and root inside. Together, the cotyledons, baby leaf, and root make up the seed. The cotyledons are like a 'sack lunch' for the baby plant. They provide food for the tiny plant until it can grow leaves big enough to feed itself!

If you grow a bean plant in some dirt in a plastic cup, you can actually watch the cotyledons shrink a little every day as the growing plant uses up the food in them. Then, when the leaves get big enough, what is left of the cotyledons falls off. Even when they fall off, they continue to feed the little plant, because what is left of their stored food goes into the soil next to the plant. There, it is taken up by the roots.

Other types of seed, like corn, only have one cotyledon; they usually send up leaves right away and start making their own food sooner than plants with two cotyledons. Because of this, their cotyledon gets 'eaten up' by the growing plant, until there is no more seed left. By the time the cotyledon is gone, the little plant is already making its own food. God knows exactly how much food to 'pack' for His baby plants!

"In all created things discern the providence and wisdom of God, and in all things give Him thanks."
—St. Teresa of Jesus

Questions

1. What provides food for the baby plant until it has leaves?

   _____

   _____

2. Along with water and sunlight, what do leaves use to make food?

   _____

   _____

Experiment!

Would you like to do an experiment with seeds?

Plant three or four bean seeds in a plastic cup. [If you use a clear plastic cup, you can spot some of the roots as the plants grow. Set the clear plastic cup inside a colored cup to keep it nice and dark for the growing roots.] Plant three or four corn seeds in another cup. Label the cups and write the date of planting on the side. Water the seeds and then make a little 'tent' with a plastic bag over the cups. Set the cups in a sunny window for six days. Then check the cups every day to see what changes take place.

1. How many days did it take the corn seeds to sprout? _____

2. How many days did it take the beans to sprout? _____

3. Which plant sent up leaves first? _____

4. Was it the plant with one cotyledon, or two? _____

5. How long was it, after the bean plant first sprouted, before the first true leaves appeared? [Be careful not to mistake the cotyledons for leaves.] _____

When your plants are about four inches tall, pull up one bean plant and one corn plant. Compare the corn and bean plants.

6. Is there any cotyledon left in the roots of the corn plant? _____

7. Is there a cotyledon in the roots of the bean plant? _____

8. Are there cotyledons left on the bean stalk? _____

9. Are the bean cotyledons shriveled? _____

Botany
Seeds

Traveling Seeds

Do you like surprises? A few days ago, we found a surprise in our flower bed. Growing there among the bright blue flowers was a tiny fir tree no bigger than my thumb. I don't know who planted it there, but I suspect the 'winged' fir seed was dropped by a chickadee and carried by the wind to the welcoming soil of the flower bed.

Are there any chickadees where you live? It is delightful to watch them in the fall, when the fir cones open to drop their seeds. The chickadees hang almost upside-down from the cones, probably picking both insects and seeds from the cones. When seeds are dropped, a single, tiny 'wing' causes the seed to drift and fall some distance from the tree. There the little seed may start to grow into a whole new tree.

Since we rely so much on plants for our food and for other uses, it is good that seeds can spread. Once again we see God's creativity in the variety of ways that He designed seeds to 'spread out,' starting even more plants to provide for both our needs and enjoyment.

One of my favorite seed designs is 'flying' seeds. Do you like to blow dandelion puffs, too? I like to watch the wind carry the little parachute-like seeds away, out of sight. Thistles and cattails also have parachute-like seeds. Although they have no 'parachutes,' maple seeds

Science 2 for Little Folks

have 'wings,' somewhat like fir seeds, that carry them to new places on the wind.

Then there are the seeds that like to hitch a ride on a passing animal or on your clothing. These seeds, called *burrs*, are usually rough and scratchy to help them stick to whomever is giving them a ride. They can travel for many miles before they fall off, all ready to grow many more burrs in their new home.

Seeds that probably hold the record for the greatest distance traveled would include those that float. Seeds blow, roll, or wash into creeks and rivers and then can float for miles downstream until they wash up on muddy banks. There, they sprout and grow. Hollow coconuts have been found bobbing merrily in the ocean, thousands of miles from land, headed for a new life on distant islands!

Some seeds don't travel at all, such as those inside certain fruits and vegetables. Think of the size of a pumpkin, which certainly isn't going to blow away or stick to a passing dog! Have you guessed that God thought of a different way for some seeds to move to a new place? Seeds that come from vines don't move; the vines do!

Squash and pumpkin vines often grow twenty feet long in just one summer. Snug inside the squash at the ends of those vines, the seeds wait out the winter weather. When spring comes, the squash softens and leaves its seeds on the ground to start a new plant.

Juicy grapes and blackberries, too, grow from spreading vines and canes. What began as one small plant can travel and bring forth new plants that bear even more delicious fruit for all to enjoy.

Jesus once told a story about seed that fell onto good soil, grew to maturity, and then produced one hundred times more seed. [Mk. 4:8] He was really telling His listeners not about seed, but about accepting His teachings and then spreading the Faith to others. Like seeds that travel, bring forth new plants, and bear delicious fruit, Jesus calls us to carry our Holy Faith to those outside of His family. With God's help, we can plants seeds of faith. Then we pray that our Holy Faith will spread in hearts throughout the world, bearing the sweet and eternal fruit of salvation.

"We marvel at miracles, not so much because of their wonder but because of their infrequency. It is a wondrous thing that seed brings forth the harvest of wheat to feed all mankind; but it

happens year after year...It is not that we should wonder less at the miraculous, but that we should come awake to the marvel of the ordinary, and wonder most of all at the Lord of the universe..."
—FROM *MY WAY OF LIFE*, BY WALTER FARRELL, O.P.

Questions

1. Name a plant that has 'flying' seeds.

   _____

   _____

2. What is the name for rough, scratchy seeds that stick to animals or clothing?

   _____

   _____

3. Name a plant whose seeds grow on long vines or canes.

   _____

   _____

Something to Do

Collect as many different types of seeds as you can find. A good time to begin your collection is in the late summer or early fall. Can you tell how each of the seeds travel? Can you identify and group all the flying seeds together? Can you group the rest of the seeds according to the way that they are spread?

If you like discovering new things about plants, you might find *botany*, the study of plants, in the encyclopedia. You may also look up the famous *botanist*, Gregor Mendel. Even greater than his *botanical* work were the 'seeds of faith' that he planted, for Gregor Mendel was, first and foremost, a priest.

Science 2 for Little Folks

Biology

Entomology: Butterflies and moths

So, What's the Difference?

Can you tell the difference between a moth and a butterfly? Because God made so many kinds of moths and butterflies, it can sometimes be difficult to tell them apart. We often think of butterflies as being more colorful, but some butterflies are very dull, and some moths are quite pretty. You can't always tell a 'critter' by its clothes, can you?

Butterflies and moths are alike in many ways. Both are covered with tiny, powdery scales. [I'll bet that scales make you think of fish, snakes, and lizards—not butterflies. Doesn't God make wonderful surprises for us to discover?] These scales cover their wings and every bit of their bodies; if the delicate scales are rubbed off their wings, neither butterfly nor moth can fly.

Both moths and butterflies have four wings, and both have 'feelers,' or **antennae**, growing from their heads. But these flying insects are different in many ways, too. Moths' antennae are usually fuzzy or feathery; butterflies have thin, thread-like antennae with a little knob at the very top.

Other differences between the two insects are found in their body shapes and in their 'busy times.' Moths tend to have thick bodies, while butterflies have slim bodies. Moths are busy at night; butterflies are out and about during the day.

When you first see a winged insect land on a flower, you might think 'butterfly'! Or, you might be scared by what looks like a wasp and take off at a run. You wouldn't want to catch a wasp, or you might get stung! But did you know that there are even a few harmless moths that look like wasps? If you'd like to start a butterfly and moth collection, knowing a bit about each flying insect will help you to *discern,* or tell the difference, between insects. Judging 'which is which' might even keep you from being stung!

Choosing our friends wisely is a good idea, too. Judging between others' good and bad behavior and speech might keep you from being 'stung' by sin!

Now, you may have heard someone say, "But the Bible says, 'Judge not, and you will not be judged.'" But Holy Scripture also tells us to be wise in judging, or *discerning,* between what is holy and what is sinful. Sometimes it seems that almost everyone wants to copy the actions of a famous singer or style of clothing, without any thought to whether the actions or styles are holy or whether they are sinful. If we cannot first judge the difference between right and wrong, how can we possibly choose what is pleasing to God?

Just as you can tell the difference between moths and butterflies by comparing the way they look and act, so can you learn how wisely to choose friends who will help you get closer to Jesus. You can discern that moths are active at night and butterflies are active during the day. So, too, some people live in the darkness of sin while others are children of the light. Let us discern, then, between what is holy and what is sinful, and choose to follow the way that will keep us always in the Sonshine.

"For once you were darkness, but now you are light...walk as children of light [for the fruit of light is found in all that is good and right and true], and try to learn what is pleasing to the LORD. Take no part in the unfruitful works of darkness, but instead expose them."

—Eph. 5:8-11

"Do not conform yourself to this age but be transformed by the renewal of your mind, that you may discern what is the will of God, what is good and pleasing and perfect."

—Rom. 12:2

Science 2 *for Little Folks* **81**

Questions

1. Name two ways that moths and butterflies are alike.

   _____

   _____

2. Name two ways that moths and butterflies are different.

   _____

   _____

3. What is another name for the 'feelers' that grow from a moth's or butterfly's head?

   _____

   _____

Experiment!

Would you like to catch a few butterflies and moths and compare them? Let's do it!

You know that moths usually are active at night. The easiest way to catch them is to put a lamp right against a low window, and the moths will fly to the light. There, you can catch them with a net.

Butterflies are out and about during the day. They can often be found feeding at purple and pink flowers. Try to handle the insects as little as possible when you catch them. By being careful in the way that you handle them, the scales will stay in place on their wings and they will still be able to fly when you are finished looking at them.

Place your insects in a clear plastic jar or glass. Make sure that the cover has a few air holes. Use a magnifying glass to examine your 'critters.'

- Did you catch the insect at night, or during the day? Does the insect have a thick body, or a thin body?
- Does your insect have little 'knobs' at the end of thin, threadlike antennae? If so, it is probably a butterfly.
- Does your insect have fuzzy or feathery antennae, with no 'knobs' at the end? If so, it is probably a moth.

Biology

Entomology: Moths

Tasty Moth or Scary Owl?

Have you ever been surprised, as you passed a pebble lying on the sidewalk, to discover that it was really an old penny? You had nearly left it there on the sidewalk, thinking it was just a pebble, but that second look made you just a little bit richer!

Sometimes birds are surprised by second looks, too. At night while you are sleeping, flying insects called *moths* are out and about. Some are in search of food; many drink their supper of fresh flower nectar through a long, hollow tongue called a **proboscis**. Like butterflies and bees, when moths gather nectar, they also carry pollen from one flower to another. This carrying of pollen from one flower to another, or *pollination,* makes the flowers grow fruit or seeds which, in turn, make more plants. God designed some plants so that they bloom only at night, when most bees and butterflies have stopped their work and gone to bed. If the moths didn't pollinate the night-blooming flowers, they might not get pollinated at all, and then the plant would die out forever! So the moth plays an important part in making sure that those plants keep on growing.

Well, what about the birds being surprised by second looks?

When moths are at rest, they often close their wings. Some moths, with their wings closed, look a little like plain, brown butterflies. A bird might

Science 2 *for Little Folks* 83

see them and think, 'Aha! A tasty snack!' But, when the moths see danger coming, they flip open their wings to show the big, glaring eyes of an owl! Yikes! The hungry attacker takes a second look and flies off in a hurry, leaving the moths safe and untouched. God has decorated the wings of these moths with *eyespots*. These eyespots make the moths appear to be something else, something much greater and more powerful than the little insects that they really are.

That makes me think of wanting to follow Jesus so closely that we begin to 'look' like Him. As we grow more and more holy, we are filled more and more with God's *grace*, or His life within us. His grace allows us to do greater and more powerful things than we could ever do without Him.

When we share a kind word or helping hand, others go away 'richer,' because they have 'picked up' a little of God's love. And if our words and actions make others take a second look at us, wouldn't it be fine if they saw only Jesus?

"If I love Jesus, I ought to resemble Him."
—St. Peter Eymard

"When You are our strength, it is strength indeed, but when our strength is our own, it is only weakness."
—St. Augustine, Confessions, 4, 16

Questions

1. What is another name for the moth's hollow tongue?

   _____

   _____

2. The act of carrying pollen from one flower to another to cause fruit and seeds to grow is called:

   _____

   _____

3. Another name for God's life within us is:

   _____

   _____

Biology
Annelids

It's a Dirty Job

Do you like the flavor of old, damp coffee grounds? How about rotting leaves, raw potato peels, and crushed egg shells? No?

To an earthworm, that's the menu for a mouth-watering feast. By eating such a 'feast,' earthworms change the garbage around them into rich, useful soil. Rich soil makes for strong, healthy plants. Healthy plants, in turn, grow more healthful food for us to eat. Does this sound a bit like *recycling,* or a fine way to use something a second time so that it isn't wasted? People didn't invent recycling; God did, and worms are His wiggly 'recyclers.'

These wiggly critters also 'bust up' tough soil. If soil is too 'heavy,' plants cannot sink their roots far into the ground. With only a few, short roots in the hard dirt, plants won't grow well. But as earthworms eat rotting leaves and garbage, they also dig little tunnels into the hard earth. These tunnels loosen the soil and let in more water for plants to 'drink.' In the loose, moist soil, roots grow better and plants get stronger and bigger. Earthworms, then, are helpful to both plants and soil.

But how do worms dig tunnels? They sure don't have itty-bitty shovels; worms don't even have arms or legs! No, they 'suck' dirt and garbage into their mouths, 'plowing' tunnels as they eat their way through the soil. With no teeth for chewing, the worms instead eat tiny bits of gravel that help mix the dirt and leaves that the worm has eaten into a soft, rich 'dirt paste' called *castings.*

Science 2 *for Little Folks* **85**

Worms often push the castings into little piles that you can see on top of the ground.

While you can see their castings, you won't see many worms above ground. But there may be as many as half a million worms living in your back yard! Those worms can 'recycle' and 'plow' several truckloads of garbage and dirt every year. It's a big, dirty job, but those simple little worms keep at it day after day after day. Their steady work cleans up the neighborhood by 'recycling' all that trash and hard dirt, changing it into useful soil.

Take a look, and I'll bet you can see 'garbage' around you, too. Maybe it is the 'trash' of a messy room that needs to be picked up. Perhaps it is the 'dirt' of sin in the culture around us, or even a 'trashy' attitude. What a difference we can make if we are willing to 'eat a little dirt' to clean it all up. With God's help, we can do it! Like the simple earthworm, we just need to keep at the job and not give up. The joys of Heaven are worth the effort.

"Let us not grow tired of doing good, for in due time we shall reap our harvest, if we do not give up."

—Eph. 6:9

Questions

1. A fine way to use something a second time so it is not wasted is called:

   _____

   _____

2. How many teeth do worms have?

   _____

   _____

3. What is another word for the rich 'dirt paste' that earthworms make?

   _____

   _____

4. What can you 'clean up' today? List at least two things.

   _____

   _____

Biology

Habitats and ecosystems

Life-giving Water

Long ago, a hungry raccoon lumbered across a dry, grassy field. He was headed for a creek some distance away, where he planned to find lunch to fill his growling tummy. At the creek, he would turn over mossy rocks in search of pinching crayfish hiding underneath, and stretch his paws into muddy holes where tadpoles wriggled. There was no reason to look for food in the field for, without water, it seemed nearly lifeless but for the grass.

Then, one day, there was a violent shaking of the ground. An earthquake split the grassy field creating a thin crack in the earth. The crack zig-zagged all the way from the bubbling creek to the dry field. In short order, the field was no longer dry! Water moved along the jagged crack that the earthquake had torn. The dry and lifeless field was becoming a shallow pond, or marsh.

Years passed. Mosquitos made their home in puddles at the edge of the marshy pond. Through the little crack that was now a tiny stream, small fish had found their way from the creek into the pond. Bug-eyed frogs made themselves comfortable in the cattails and on lily pads at the water's edge. The lifeless field of old was no longer lifeless at all!

Mosquitos whined low over the marsh, to be gobbled up by leaping fish and long-tongued frogs. And now that there was plenty of water, trees grew over and shaded the pond's border. In the tree branches above the water, blue and white kingfisher birds perched, waiting for fish to swim by in the leafy shadows below. If you had been there watching, you might have seen a kingfisher dive swiftly from his perch into the pond. When the bird came dripping out of the water, in his sharp beak was a silvery fish!

Bandit-masked raccoons no longer had to hike all the way to the creek for lunch. Their tracks could be found wandering about the marsh, stopping here and there where the animal's small black fingers had snatched a bullfrog or slippery minnow from the shallow water.

Kingfishers were not the only birds in this marshy home, or *habitat,* of water-loving plants and creatures. Long-legged herons and colorful ducks came to dine on the tadpoles, fish, and frogs. Some of the ducks stayed, nesting in the pond's thick clumps of bulrushes. In their nests, they laid so many eggs that sneaky raccoons now and then snatched some for a snack. [Would you like to snack on raw duck eggs?]

Can you see that each thing in this watery home was in some way linked to another? For example, the mosquitos became food for the frogs and fish. The frogs and fish then became food for the larger animals and birds. Even the trees and bulrushes were an important part of this home, giving shelter to the animals living in the marshy habitat. All of these things that God created were connected in some way to one another. A connected group of living things, along with their habitat, is called an *ecosystem.*

But if there had been no water, there would have been no whining mosquitos, nor croaking frogs, nor silvery fish, nor diving kingfishers, nor bandit-faced raccoons leaving their hand and footprints in the soft mud between the cattails. Without life-giving water, there could have been no life-filled, marshy ecosystem.

When you think of life-giving water, I wonder if you also think of a sacrament. Do you think of Holy Baptism? Before we were baptized, our souls were spiritually dry and lifeless. But when the waters of Baptism touched us, a miraculous change took place. Our souls were flooded with God's grace! In an instant, we became connected to a great, living family, linked to one another by a sharing in the life of our Heavenly Father. Members of this family are joined across all nations, and even across time and space, for we are linked to all the saints living in Heaven as well.

But Baptism is only the first of many sacramental connections that link us to Heaven. Next comes the Sacrament of Reconciliation, followed by the Sacrament of Holy Communion, then Confirmation, and so on. Each of these sacraments makes us more and more alive in Jesus Christ, preparing us for the never-ending excitement of Heaven. But without the holy waters of Baptism, *none of this life would be ours.* Say, have you thanked God—and your parents—for the gift of Baptism?

"But when the kindness and love of God our Savior appeared, He saved us... through the baptism of new birth..."
—Titus 3:4 & 5

"O God, You are my God whom I seek; for You my flesh pines and my soul thirsts, like the earth, parched, lifeless and without water."
—Ps. 63:2

Questions

1. Name some living things that you might find in a marshy habitat.

   _____

   _____

   _____

   _____

2. A group of living things, along with their habitat, is called what?

   _____

   _____

3. Name the first sacrament that links us to Heaven.

   _____

   _____

Science 2 *for Little Folks*

Biology
Migration, hibernation, and estivation

In Wisdom You Have Made Them All

Would you rather be outside on a steaming hot day, or on a snowy, frozen day?

I once knew a Montana cowboy who cared for his snow-crusted cattle in the middle of a blizzard, when temperatures were below zero [-18º C]. This cowboy also saddled up behind a dusty, thirsty herd on August days when the thermometer read 100º F [38º C]. Someone asked him if he liked the hot days better, or the frozen days better. He answered that he preferred cold weather, because he could always put more clothes on, but he could only take so many clothes off.

Perhaps you live in or have visited a southwestern desert climate, where summers are long and very hot, but winters are short and mild. Or maybe you live in or have visited a northern boreal forest, where winters are long and extremely cold, but summers are short and mild. Do you ever wonder how the plants and animals that live in those places can survive, or live, through the harsh climate? Desert critters don't have air conditioning or lemonade to make them comfortable in the heat. Northern critters can't put on a heavy coat and an extra pair of mittens before they go outside, or enjoy a mug of hot chocolate with marshmallows when they come inside. Let's see how, in His wisdom, Our Lord has provided for the plants and animals in climates with very high or very low temperatures.

Many animals call the boreal forests of Canada their home. Squirrels, garter snakes, Canada Geese, and Snowshoe Hares are just a few of the animals that live in this cold ecosystem, where temperatures often fall to -40° F [also -40° C]. Nevertheless, God planned unusual ways to keep His animals warm and protected.

Some of the animals, like the squirrels and snakes, spend the winter *hibernating,* or going into a kind of deep sleep, in their underground dens far below the snow. Then, when temperatures rise in the spring, they come out of their dens again to stretch in the warm sunshine. Other animals, like the Snowshoe Hare, stay awake all winter but have extra warm coats and wide, furry feet. These specially designed feet make it easy for them to hop across the top of deep snow, where you and I might stumble and fall. Still other critters, like the Canada Goose, *migrate,* or move to a warmer climate during the coldest part of the year. Then, just like people who have taken winter vacations down south, they fly back to their northern home when the weather warms up.

God has even more clever designs for His boreal forest creatures. Did you know that most snakes, in warm climates, lay eggs? After a time of being warmed by the sun, baby snakes hatch from the eggs. But during the short northern summers, there are not enough warm days to help snake eggs hatch. So Our Lord designed these boreal garter snakes to bear their babies inside their bodies, instead of hatching from eggs!

Our Lord even designed plants for cold habitats and ecosystems. Boreal forests are full of evergreen trees like pine, spruce, and fir. Most evergreen trees are shaped a little like a tall, pointy triangle. This shape makes heavy snow slide off the tree branches instead of weighing them down and causing the branches to break. All the sliding snow piles up to make cozy little 'snow caves' under the tree where animals can find shelter. Even better, most evergreens have cones stuffed with rich seeds. Squirrels and other hibernating critters eat the seeds in the summer and grow fat, so they can go to bed for the winter with full tummies.

Now, if I tell you that roasty deserts also have hares and snakes and squirrels and birds, you will probably wonder how animals that live so happily in the cold north can also live happily in the hot southwest. The answer is that their Creator made the desert hares and snakes and squirrels and birds just a little different from their northern relatives—perfect for desert life.

While some northern animals sleep through the worst of the winter, some desert critters [such as squirrels] instead sleep through the worst of the summer! This long summer sleep is called **estivation**. Other creatures escape daytime temperatures that may reach 120º F [49º C] by staying in the shade or moving underground all day. Then they come out at night, when the temperature falls. [Animals that are active at night are called *nocturnal* animals.] Still other creatures, like the rattlesnake, are active only around sunrise and sunset, when it is still light outdoors, but not so hot as in the daytime.

Just as our Creator has clever designs to make life easier for critters in cold temperatures, He has smart ideas for His desert critters, too. For example, the desert 'floor' is terribly hot. So many birds who live there, like the roadrunner, have extra long legs to keep their bodies a little farther away from the heat. You remember that the Canada Goose migrates south to get away from the northern cold. When summer comes to the desert, some desert birds migrate north to get away from the heat! And you may also have guessed that many desert snakes, like the Mojave Shovelnose Snake, do lay eggs since the warm soil can help the eggs to hatch. Doesn't God have good ideas to keep animals comfortable in their very own habitats?

Even desert plants were created just for their hot, dry climate. For example, the Barrel Cactus really does look and act a bit like a water barrel. It has no branches, and no leaves; it looks sort of like a stumpy, prickly barrel. But the unusual roots of this cactus spread far and wide in a big circle, to catch as much rain water as possible. Sometimes it may rain only once or twice a year in the desert, so the cactus has to fill up! This plant is so good at filling up that it can live for several years with the water that it stores from just one rainfall.

Do you wonder, if it only rains a few times a year, how the animals can find enough water so they don't die of thirst? If you guessed that God made a plan for that, too, you were right! Most animals who live in the desert have bodies designed to use water extra carefully. Like the Black-tailed Jackrabbit [that is really a hare], many get all the water they need by eating water-rich cacti or other plants.

Now, I have some questions for you! If all hares [or rabbits], seem to be pretty much the same critter from their long ears to their bunny tails, how did one get to be so well suited for the desert, and one for the frozen north? Do you think that the Snowshoe Hare woke up one day and said to himself, 'I think if my feet were wider and furrier, I could walk on snow better, so I am going to grow a different type of foot for myself today.' Or do you think, perhaps, that a shovel-nose snake decided one afternoon that it wanted to lay eggs instead of bearing live babies? No, these were not changes that the animals chose to make so that life would be easier for them in their different habitats.

Rather, it was their Maker, our Lord Jesus Christ, Who planned the differences in these animals and plants. In God's own way and time, He made these creatures to fit perfectly in their ecosystems. He gave them all that they needed to live and grow happily right where He put them.

Just as Our Lord has a perfect plan for His perfectly designed animals, He also has a perfect plan and design for you. Oh, it doesn't mean that each day is easy, or that you always have just what you want. Even the northern squirrels have to work hard to gather seeds and nuts to eat before they can rest for the winter! But God created you to live and work in a family 'habitat,' within a community and Church 'ecosystem.' There is even a special name for this 'ecosystem.' It is called the Kingdom of God and, because God made you His child at Baptism, you are already a precious member. *You are a precious part of God's loving plan.*

"The eyes of the Lord are upon those who love Him, a mighty protection and strong support, a shelter from the hot wind and a shade from noonday sun, a guard against stumbling and a defense against falling."

—S<small>IR</small>. 34:16

"O LORD, how manifold are Thy works! In wisdom hast Thou made them all..."

—P<small>S</small>. 104:24

Questions

1. Animals who go into a kind of deep sleep in the winter are said to be doing what?

_____

_____

2. When a bird flies to a different climate to find better weather, it is said to do what?

_____

_____

3. 'Sleeping' through the summer to avoid the heat is called what?

_____

_____

4. What word describes animals that are active at night?

_____

Environmental Stewardship

The Paper Bag Prince

Dominion Over the Earth

Have you ever wanted to sit on the couch, but someone was already sitting there and said you had to go sit somewhere else because there wasn't enough room? Yet, perhaps another time your Grandpa was sitting on the same couch reading, and there was enough room for three or four more children to snuggle against him to hear the story. How can a couch that was too crowded for two people be perfectly comfortable with four or five? It wasn't the 'room' on the *couch*, but the 'room' in the *hearts* of those sitting upon it. A sharing heart makes all the difference!

When Our Lord created the vast, beautiful earth, He commanded that His children, "Be fruitful and multiply, and fill the earth..." [Gen. 1:28]. Do you think that the world is big enough to hold everybody that God creates? Some people think that the earth already has too big a *population,* which is the number of people that live in a place. Is it possible for the earth to be too full, or is it maybe just that people often don't want to share?

Some people think that there won't be enough food to feed all the children that God will create. Do you suppose that God makes mistakes? Could He have forgotten to plan for ways to feed everybody? NO! We know that God *never, never* makes mistakes.

God knows all things, so His plans are always perfect. You have learned that science is a way of studying and discovering what God already knows, because He made it all in the first place. But science should also include using what we find to benefit ourselves and others, to the greater glory of God. He created for our use the earth's *resources*, or useful things like minerals, water, and plants. If we seek God's inspiration, use our intellects, and the earth's resources, we will find that there is plenty for all of God's children.

For example, farmers of long ago did not know how to get water to thirsty plants. If it did not rain, the plants died and there was no food to eat that year. Families went hungry, not because there were too many people, but because there was not enough food. In those ancient times, farmers also had no tractors for plowing or machines to harvest their crops. It was difficult for a farmer to feed even his own family. Then, as time passed, people thought of new ways to *irrigate*, or bring water to crops. They invented machines that made it easy for one farmer to grow plenty of food not only for his own family, but for hundreds of other families, too. As Our Lord blessed His earth with more precious children to know, love, and serve Him, He also showed us how to grow more food than ever to feed them.

God also commanded that we *subdue* and have *dominion* over the earth. The earth did not make us. It is not our 'mother,' nor is it somehow more important than the people whom God created to live upon it. To subdue and have dominion over the earth means that mankind is to bring it under control and have authority over it.

If you ever visit Israel, you can see fine examples of 'taking dominion over the earth.' *Engineers* and *agronomists* have developed ways to grow huge, vegetable-filled gardens in the harsh desert, using very little water and special types of plants.

Scientists continue to discover more and more ways that Our Lord provided for us. When He created the earth, He planned for our every need. If we are willing to cooperate with His plans, He will help us learn how to find and use all the resources that He made. Perhaps, with the inspiration of the Holy Spirit, you will someday be able to use what you learn from studying science to respond to God's command: "fill the earth and subdue it."

"All that is best, from Thee comes down
On us, with rich and ample store..."
—FROM 'SURSUM CORDA,' [19TH CENTURY, ANONYMOUS]

Questions

1. What is another word for the number of people that live in a place?

2. Does God ever make mistakes?

3. Is the earth our mother?

4. Find the definitions of engineer and agronomist in the dictionary.

Something to Do

Is it possible to plant a garden on a steep mountain? Farmers used to say that it was not, because watering washed both plants and dirt down the mountain side. Then farmers learned to make terraces, which are like wide steps, on the mountainsides. Using terraces, farmers can grow more crops in mountainous countries around the world. Look up *China*, *Japan*, and the *Philippines* in an encyclopedia for pictures of these terraces.

Perhaps you can experiment with terraces if you have a place for digging in your yard and the permission of your parents.

Biology
Fetology

Angel In The Waters

Before You Were Born

Aren't babies wonderful? Oh, sure, they can be a bit noisy and smelly at times, but their baby smiles and giggles are loving reminders of the One Who sent them.

We usually think that the day of a baby's birth is the day that its life begins. Of course, the most important 'birth' that the baby will have is at its Holy Baptism, when the effect of original sin is washed away, and that tiny person becomes a pure and grace-filled child of the King! But do you realize that a baby's life really begins long before its birth? Oh, yes, that newborn had been kicking up its heels and turning somersaults, happily hidden away in its mother's womb where no one could see how much fun it was having. Let's go back to the beginning.

At **conception**, the moment the mother's seed is joined with the father's, God makes a brand new person, in His image, with its very own *immortal soul*. God made only one of you, with nobody else ever before or ever again like you. God made you *unique*, for a special purpose the He gave to no one else. It is the same with the new baby. He or she is also unique in history and in the sight of God. [Can you find 'unique' in the dictionary?] Then God tenderly began to form the new baby in its mother's *womb*, which is the special place near the

98 *Science 2 for Little Folks*

mother's tummy and not far below her heart, that God made just for babies.

Only a few weeks after the baby's conception, when it is still so small that you would have a hard time seeing it, its little baby heart begins to beat, sending its own blood cells through its own tiny blood vessels. Everything that will decide what color eyes and hair and skin this tiny infant will have is already part of the baby. Warm and protected, this tiny person floats and 'swims' inside the womb, growing from one cell at conception to over *sixty trillion* cells at birth!

Just as you started life as one cell, so too at the Annunciation did Jesus begin His Incarnate life. What dignity Our Lord gave to the unborn, by choosing to start His earthly life as a tiny Baby! Even though He had just begun His hidden life in the womb of His Virgin Mother, how joyously His cousin St. John the Baptist greeted Him and Our Lady when they arrived to visit St. Elizabeth. Do you remember the most amazing part of the story? Not only did St. John recognize our Infant Lord, but St. John was an infant in his mother's womb at the time, also!

It is important to remember that, even though the baby doesn't look much like a baby for the first few weeks of its life, it is still a living person with a living soul, *just like you were at the same age*. Just because we can't see with the eyes and understanding of God, it doesn't mean that the baby is less important to Him or to us. If you gave a toddler a choice between a cookie and a $50 bill, which do you think he would choose? If he did not take the money, would it mean that the money wasn't worth anything? No, it would just mean that the toddler didn't have enough understanding to know how much the money was worth. Some people don't know the value of tiny babies still in the womb, because they don't see with God's understanding.

As the precious tiny person grows in the womb, around the sixth week of pregnancy, its tiny teeth buds form in its baby gums. By two and a half months, the baby has fingerprints and all his organs, like his heart and lungs, are in place and doing their jobs.

As the unborn child keeps on growing inside his mother, he can hear sounds, like his parents' voices. He can even see light if it is flashed on his mother's tummy. He is breathing the liquid in the womb to exercise his lungs. [If you or I did that we would drown, but our wonderful Maker made the unborn baby to get his 'air' through the **umbilical cord**. The umbilical cord, which is a little bit like a long hose, also sends 'food' to the baby until he is born. Your umbilical cord used to be attached where your belly button is now!] The baby also sucks his thumb, sleeps, hiccups, and kicks, all while living inside his mother. What a great life!

Psalm 139 says it this way: "Truly You have formed my inmost being. You knit me in my mother's womb. I give You thanks that I am fearfully, wonderfully made; wonderful are Your works."

Questions

1. The moment the mother's seed is joined with the father's is called:

_____

_____

2. The special place near the mother's tummy that God made just for babies is called:

_____

_____

3. What is the name for the 'hose' that delivers 'air' and 'food' to the baby before it is born?

_____

_____

Something to Do

First, find and read the beautiful story of the Visitation in Luke 1: 39-56. Take a little time to think about verse 43: "...who am I that the mother of my Lord should come to me?" Perhaps you can write the verse on a pretty card and place it next to a picture or statue of the Blessed Mother in your home.

Now, ask your mom or dad if you may see their baby pictures. Have their looks or size changed since they were babies? If a person's size and looks have changed since babyhood, does that mean that he is not the same person to whom God gave the gift of life at conception?

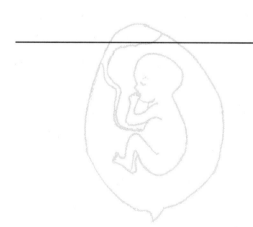

Biology

Infant nutrition

Baby Food

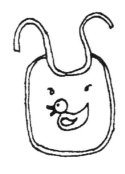

Do you have special birthday customs at your house? At our house, the 'Birthday Child' gets to pick whatever she'd like for dinner. Sometimes our menus look like this: hot corn on the cob, fluffy mashed potatoes and gravy, crispy chicken, and chocolate ice cream. Does that sound good to you? If it is a good 'birth-day' meal, would it be a good meal for a baby who had just been born? No, a new baby could not eat that meal, because its tiny body needs only one thing: milk!

God's perfect food for babies is milk from their mothers. When a mother *nurses*, or feeds her baby from her breast, she is giving it the best food possible. Mother's milk not only makes her child grow big and strong, but helps keep the baby healthy, too. That is because there is a special ingredient in the milk called **antibodies**. Antibodies are disease-fighters that give the baby extra protection against germs.

Most people think that, when a mother nurses her baby, it is good just for the baby. But Our Lord, in His always perfect plans, made nursing benefit the mother, too. Women who nurse their babies have more protection against some types of cancer. A nursing mother also recovers more quickly from childbirth. Isn't God good to think of *everything*?

Most babies spend a full nine months in the womb before they are finished growing and ready to be born.

Sometimes, however, babies can be born too early, at eight or seven or six months of pregnancy. If you have read the story 'Before You Were Born,' you already know a little about the life of an unborn baby.

Long before a baby is born, it is practicing breathing, and sucking, and all of its little body is working at growing. However, not every part of the baby is *mature*, or ready to work on its own. Even a healthy newborn is not mature enough to take care of all its own needs. It has to have a mom and dad to feed and care for it.

A baby born prematurely [before it is mature, or ready to be born] is called a *premature infant,* or 'preemie.' These babies did not have time to finish growing before they were born. Their tummies were not quite ready for food, and their lungs weren't quite ready to do the work of breathing. What would happen if you could not eat or breathe properly? You would be very sick, wouldn't you? That is the worry with preemies. If they cannot eat or breathe well, they can become very sick. But guess what—*God has a special plan for preemies.*

In the past, doctors had a hard time making preemies well. But in the last several years, scientists have made many discoveries of God's special care for His tiniest children. One of these discoveries is that, when a woman gives birth to a premature baby, her milk is different than it would be if the baby had been full-term, or in the womb for the normal nine months!

Can you figure out how God might make the mother's milk different for a preemie? If a baby has trouble digesting its food and difficulty breathing, it would be pretty important to work at fixing those problems, wouldn't it? Well, that's just what Our Lord did. Scientists have discovered that women who give birth prematurely have milk that is even richer in 'growth factors' that speed up the maturing of the baby's lungs and digestive system.

The premature infant also needs a boost to grow the strong bones and muscles that he would have developed within the womb. So God gave this specially designed mother's milk extra *protein* for muscles, *calcium* for bones, and antibodies to fight germs! Finally, scientists have learned that this special milk continues to be made in the mother's body for several weeks after the baby is born, to give the infant the extra help that it needs. How well Our Lord provides for us!

For all His wondrous works, may Jesus Christ forever be praised!

*"She of the King of Stars beloved,
stainless, undefiled,
Christ chose as His Mother-nurse,
to Him, the stainless Child;
Within her breast, as in a nest,
the Paraclete reposes,
Lily among fairest flowers,
Rose amid red roses."*
—from 'Hymn to the Virgin Mary,' by Conal O'Riordan

Questions

1. What are the disease-fighters that give protection against germs called?

_____

_____

2. Look up *calcium* in the encyclopedia. Why do we need calcium in our diet? Which foods are rich in calcium?

_____

_____

_____

_____

3. Look up *protein* in the encyclopedia. Why do we need protein in our diet? Which foods are rich in protein?

_____

_____

_____

_____

Biology

God's precious children

Power in Weakness

When I was little, my daddy would sometimes lift my brother and me high above his head, with each of us seated on one of his powerful hands. I was never afraid that my daddy would drop me, for I knew how strong he was and how much he cared for me.

Have you ever wished that you were bigger and stronger so that you might be able to do more for Our Lord? Usually, as we grow bigger, we grow in ability as well. Sometimes, however, people still cannot do what they might like because of *disabilities*. Their minds or bodies do not work in the way one would expect. Yet, one of God's marvelous mysteries is that He often *especially chooses* those who are weak to do His work.

Did you know that St. Paul had a disability? Some think that he had very poor eyesight. Whatever his disability, when St. Paul prayed that Our Lord would take it away, God told him, "My grace is sufficient for you, for My Power is made perfect in weakness." St. Paul was then content with his disability, because he knew that God was the one Who would provide the power and do the work in His own way. He would even use St. Paul's disabilities to do good.

Just as there are disabilities that make it hard to see, there are disabilities that make it difficult to walk or talk.

Nerves send messages to your muscles to tell them what to do. Your brain is the 'command center' for most of those messages. But sometimes the brain can be injured when a baby is born, and the nerves are never able to send messages properly. This injury is called **cerebral palsy**. Some people with cerebral palsy have only a little trouble making their arms or legs do what they want. Others are like 'Angela,' a little girl who sometimes visits at our house. Angela is eleven years old, but she cannot talk, walk, or see. However, she smiles and laughs all the time. What do you suppose she laughs at? Do you think that maybe she is playing games with the angels? You see, 'Angela's' mind and her body are 'disabled,' but she will never commit a mortal sin. How beautiful her soul must be, and how precious in the eyes of Our Lord!

Some disabilities appear only as the body begins to grow older. Bones that were once strong become thin and break. Often, bones don't heal properly and grandmas and grandpas have to stay in a wheelchair or in bed because they can no longer walk. Some older people may have a stroke, which can cause injury to the brain. As with cerebral palsy, when the brain is injured, it can't send proper messages to the body.

Once I read of a man who had to stay in bed all the time. Because of a stroke, he could no longer walk; neither could he use his hands. He was also blind and could not see to read or write. Was this man disabled? Not in God's eyes! You may think, "What could that man do to serve God? He cannot do anything!" But that man was a great servant of God. He could pray! Day after day, people would come to the nursing home where he lived and tell him all their prayer intentions. Then he would pray with all his heart. Jesus worked wonderful miracles through this man.

It is good to study how God made our bodies and to learn how to stay healthy. Thanks be to God that He is showing us, through the science of medicine, new ways of making people well, too. But, as the *Catechism of the Catholic Church* [1508] teaches, we are not always healed. It is then that we can say 'YES' to God, and allow Him to use our weakness for His glory, just as He did with St. Paul. It is a beautiful truth of our Holy Faith that our sufferings can be united with those of Our Lord in His Passion, to help bring souls to salvation.

Thanks be to God that, even in our weakness and littleness, with all our abilities and disabilities, He holds us tenderly in His strong hands. How good He is! For, when we offer our weakness to Him, He will use it for His glory.

"...for it is You Who have accomplished all we have done."
—Is. 26:12

"Suffering...acquires new meaning; it becomes a participation in the saving work of Jesus."
—CCC 1521

Questions

1. Name a famous saint who had a disability.

   _____

   _____

2. A stroke causes injury to what part of the body?

   _____

   _____

3. God told St. Paul, "My power is made perfect in ___."

   _____

   _____

Something to Do

Do you know someone who has to stay in bed most of the time? If they already know, love, and serve Our Lord, perhaps you can visit and pray with these mighty servants of God. If they do not yet know how much Our Lord loves them, you can show His love by your presence and prayers. Visiting the sick is a Corporal Work of Mercy. If you can't think of anyone to visit, a nursing home is a good place to start!

Answer Key

p. 1. Invisible Marks
1. pheromones
2. exocrine glands
3. sweat glands
4. character

p. 4. How to Dress a Duck
1. down
2. By preening, the duck spreads oil up and down its feathers.
3. wide; no separate toes; shaped like a paddle, fan, or swim fin

p. 7. Precious Blood
1. veins, arteries and capillaries
2. red cells
3. clot
4. white cells
5. definitions will vary

p. 10. Coloring Adam and Eve
1. melanin
2. melanocytes
3. white
4. sunlight

p. 13. He Will Dry Every Tear
1. lacrimal glands
2. tear duct
3. salt and other substances

p. 16. Our Hearts and Theirs
1. definitions will vary
2. arteries
3. about 130 beats per minute
4. about 25 beats per minute

p. 19. Run for Your Life!
1. endocrine
2. hormones
3. epinephrine and norepinephrine
4. pituitary

p. 22. Eyes to See
1. iris
2. pupil
3. No. Staring at the sun can damage your eyes.

p. 25. Nerves and Your Sense of Touch
1. neurons
2. spinal cord
3. leprosy

p. 28. Ears to Hear
1. vibrations
2. sound waves
3. larynx

p. 31. Spit!
1. saliva
2. salivary glands
3. in your mouth
4. because less saliva is produced while you sleep.

p. 34. Soul Food
1. healthy
2. bodies
3. fat and carbohydrates
4. grains and legumes; especially wheat, corn, oats, rice; beans, peas and peanuts.

p. 37. Germs Make me Sick!
1. bacteria and viruses
2. cover the mouth when coughing, cover the nose with a tissue when sneezing, and stay home if one is ill.
3. moisture, warmth, and food

p. 40. Salt of the Earth
1. Our bodies must have salt if we are to stay healthy.
2. dissolved
3. minerals

p. 43. God's Building Blocks
1. an atom
2. element
3. compound
4. God

p. 46. Freezing and Solids
1. freezing

p. 48. Light from Light
1. reflected
2. 93 million miles
3. from the sun

p. 51. Falling Stars
1. meteors or shooting stars
2. about August 11-13th
3. tiny pieces of rock and dust
4. answers will vary

p. 53. Searching the Heavens
1. observatories
2. astronomers
3. solstice
4. Mount Graham, Arizona

p. 56. From the Rising of the Sun
1. no
2. east

p. 59. Merciful Rain
1. precipitation
2. water vapor
3. crystals

p. 62. Lightning!
1. negative and positive charges
2. nitrogen

p. 64. Weathering the Storm
1. weathering
2. more space
3. no

p. 68. Rock My Soul
1. volcanic rock
2. marble

p. 71. Layers of Rock, Layers of Faith
1. sedimentary
2. fossils
3. answers will vary
4. meditate

p. 74. Feeding Baby Plants
1. cotyledons
2. chlorophyll

Science 2 for Little Folks

p. 77. Traveling Seeds
1. dandelion, thistles, cattails
2. burrs
3. pumpkin, squash, blackberries

p. 80. So, What's the Difference?
1. covered with scales; have four wings; have antennae
2. Moths' antennae are fuzzy; butterflies have thin, thread-like antennae with a little knob at the top. Moths have thick bodies; butterflies have slim bodies. Moths are busy at night; butterflies are out and about during the day.
3. antennae

p. 83. Tasty Moth or Scary Owl
1. proboscis
2. pollination
3. grace

p. 85. It's a Dirty Job
1. recycling
2. none
3. castings

p. 87. Life-giving Water
1. mosquitos, frogs, birds, fish, raccoons
2. ecosystem
3. Baptism

p. 90. In Wisdom You Have Made Them All
1. hibernating
2. migrating
3. estivating or estivation
4. nocturnal

p. 95. Dominion Over the Earth
1. population
2. NO!
3. NO!
4. definitions will vary

p. 98. Before You Were Born
1. conception
2. womb
3. umbilical cord

p. 101. Baby Food
1. antibodies
2. Calcium, found in milk, milk products, and green vegetables, is essential for the growth of bones and teeth; it also aids in blood clotting.
3. Protein is found in milk, cheese, eggs, fish, and meat. Legumes, which provide incomplete proteins, can be made complete by mixing them with other incomplete or complete proteins. Protein is essential for cell growth and functioning, particularly in muscle tissue and the blood.

p. 104. Power in Weakness
1. St. Paul and many others!
2. the brain
3. weakness

Pronunciation Key

Invisible Marks, p. 1
exocrine: EX uh kruhn
pheromone: FEHR uh moan

Precious Blood, p. 7
capillaries: CAP uh lehr eez
platelet: PLAYT luht

Coloring..., p. 10
albino: al BYE noh
melanin: MEL uh nin
melanocyte: mehl uh NO site

He Will Dry..., p. 13
lacrimal: LAH kru mul

Our Hearts..., p. 16
carotid: kuh ROT id

Run for..., p. 19
adrenal: uh DREE nuhl
endocrine: EHN du kruhn
epinephrine: ehp uh NEF ruhn
exocrine: EX uh kruhn
norepinephrine: NOR ephp uh nef ruhn
pituitary: puh TOO uh teh ree

Eyes to See, p. 22
tapetum lucidum: tuh PEE tum loo SID um

Nerves..., p. 25
neuron: NYOO rahn
receptor: ree SEP tur

Ears to Hear, p. 28
larynx: LEHR inks

Spit! p. 31
salivary: SAL uh very

Salt of the Earth, p. 40
solubility: sol you BILL uh tee

Falling Stars, p. 51
friction: FRIK shun
meteors: MEE tee orz

Searching the Heavens, p. 53
observatories: uhb ZUHR vuh tor eez
solstice: SOUL stiss

Science 2 for Little Folks

Merciful Rain, p. 59
precipitation: pree sip uh TAY shun

Lightning!, p. 62
nitrogen: NYE truh jen

Rock My Soul, p. 68
metamorphic: met uh MORE fic

Layers of Rock..., p. 71
sedimentary: said uh MEN tree

Feeding Baby Plants, p. 74
chlorophyll: KLOR uh fil
cotyledon: kaht uh LEE done

So, What's the Difference?, p. 80
antennae: ann TEN uh

Tasty Moth..., p. 83
proboscis: pruh BOSS uhs

In Wisdom..., p. 90
estivation: ess tuh VAY shun

Before You Were Born, p. 98
conception: kuhn SEP shun
umbilical: uhm BILL uh kuhl

Baby Food, p. 101
antibodies: ANT uh baw deez

Power in Weakness, p. 104
cerebral palsy: suh REE brull PAWL zee

Sources

Spit! p. 31
Richard Hill, "British Researchers Have the Subject Licked: Saliva Does Heal Wounds," Research Notebook, *The Oregonian*, June 25, 1997

He Will Dry Every Tear, p. 13
Elle Becker, "The Physiology of Tears," *Mothering*, Winter 1996 n81 p.25

Baby Food, p. 101
Steven Gross, M.D., "Elevated IgA concentration in milk produced by mothers delivered of preterm infants," *The Journal of Pediatrics*, September 1981, pp.389-393

Leanna C. Read, "Growth factor concentrations and growth-promoting activity in human milk following premature birth," *Journal of Developmental Physiology*, 1985, 7, pp. 135-145

Throughout
- Scripture quotations excerpted from the *Ignatius Bible, Revised Standard Version—Catholic Edition*, © 1966 by Division of Christian Education of the National Council of the Churches of Christ in the United States of America.
- Quotations from the English translation of the *Catechism of the Catholic Church* for the United States of America copyright © 1994, United States Catholic Conference, Inc. —Libreria Editrice Vaticana.

Distributed by:
Catholic Heritage Curricula
P.O. Box 125, Twain Harte, California 95383

To request a free catalog, call toll-free: **1-800-490-7713**
Or visit online: **www.chcweb.com**

If your family enjoyed and benefited from these stories, perhaps you would also enjoy some of CHC's other homeschooling materials, whether for extra practice outside of school, homeschooling, or character development.

Other titles by Nancy Nicholson:
    Devotional Stories for Little Folks
    Devotional Stories for Little Folks, Too
    Little Folks' Letter Practice
    Little Folks' Number Practice
    Little Stories for Little Folks: Catholic Phonics Readers
    My Catholic Speller Series
    Language of God Series
    Easy As 1, 2, 3: A Catholic Overview of Science
    High School of Your Dreams

Other titles available from Catholic Heritage Curricula:
    The King of the Golden City by Mother Mary Loyola
    A Catholic How-To-Draw by Andrea Smith
    A Year with God: Celebrating the Liturgical Year
    Sewing with Saint Anne by Alice Cantrell
    Stories of the Saints: Volumes I-IV by Elaine Woodfield
    Behold and See: Beginning Science by Suchi Myjak
    And more!